Essentials of Meteorology

Essentials of Meteorology

Brady Walton

Larsen & Keller
www.larsen-keller.com

Essentials of Meteorology
Brady Walton
ISBN: 978-1-64172-469-2 (Hardback)

Larsen & Keller
Published by Larsen and Keller Education,
5 Penn Plaza,
19th Floor,
New York, NY 10001, USA

Cataloging-in-Publication Data

Essentials of meteorology / Brady Walton.
p. cm.
Includes bibliographical references and index.
ISBN 978-1-64172-469-2
1. Meteorology. 2. Atmospheric science. I. Walton, Brady.
QC861.3 .E87 2020
551.5--dc23

Printed and bound in China.

For more information regarding Larsen and Keller Education and its products, please visit the publisher's website www.larsen-keller.com

Table of Contents

Preface

Meteorology is a branch of atmospheric sciences which primarily deals with weather forecasting. Its constituents are atmospheric physics and atmospheric chemistry. The noticeable weather events which are explained using meteorological science are called meteorological phenomenon. Some of the different fields where meteorology is applied are transport, agriculture, energy production and construction. Spatially, meteorology is broadly divided into three scales - microscale, mesoscale and synoptic scale. These scales are also related directly to proportionate timescales. There are various equipment which are used within this field such as air quality sensors, flood sensors, lightning sensors and rain gauge, along with remote sensing equipment such as radar and lidar. This book provides comprehensive insights into the field of meteorology. Some of the diverse topics covered herein address the varied branches that fall under this category. Those in search of information to further their knowledge will be greatly assisted by this book.

A short introduction to every chapter is written below to provide an overview of the content of the book:

Chapter 1 - Meteorology refers to a branch of atmospheric sciences which is involved in weather forecasting. It makes extensive use of the principles from atmospheric physics and atmospheric chemistry. This is an introductory chapter which will provide with a brief introduction to all the significant aspects of meteorology.; **Chapter 2** - The observable weather events which are studied within meorology are known as meteological phenomena. A few of them are heat wave, drought, cyclone, acid rain, thunderstorm and thundersnow. The chapter closely examines these major meteorological phenomena to provide an extensive understanding of the subject.; **Chapter 3** - The branch of science which focuses on the study of the Earth's atmosphere along with the processes related to it is known as atmospheric science. The various areas of interest within this field include atmospheric layers and atmospheric composition. The topics elaborated in this chapter will help in gaining a better perspective about these areas of study within atmospheric science.; **Chapter 4** - Weather refers to the state of the atmosphere which describes the extent of how hot, cold, wet, dry, stormy, clear or cloudy it is. Applying science and technology to predict the conditions of the atmosphere at a specific location and time is called weather forecasting. The topics elaborated in this chapter will help in gaining a better perspective about the diverse aspects of weather and weather forecasting.; **Chapter 5** - There are a number of instruments and equipment which are used in meteorology such as anemometer, barometer, ceilometer, hygrometer, weather balloon, weather radar, pyranometer and rain gauge. The topics elaborated in this chapter will help in gaining a better perspective about these instruments and equipment used in meteorology.

Finally, I would like to thank my fellow scholars who gave constructive feedback and my family members who supported me at every step.

Brady Walton

1

Meteorology: An Introduction

Meteorology refers to a branch of atmospheric sciences which is involved in weather forecasting. It makes extensive use of the principles from atmospheric physics and atmospheric chemistry. This is an introductory chapter which will provide with a brief introduction to all the significant aspects of meteorology.

Meteorology is the sub discipline of atmospheric science that includes atmospheric chemistry and atmospheric physics with a major focus on weather processes and forecasting. Meteorological phenomena are observable weather events which illuminate and are explained by the science of meteorology. Those events are bound by the variables that exist in Earth's atmosphere.

Scales of Meteorology

Weather occurs at different scales of space and time. The four meteorological scales are: microscale, mesoscale, synoptic scale, and global scale. Meteorologists often focus on a specific scale in their work.

Microscale Meteorology

Microscale meteorology focuses on phenomena that range in size from a few centimeters to a few kilometers, and that have short life spans (less than a day). These phenomena affect very small geographic areas, and the temperatures and terrains of those areas.

Microscale meteorologists often study the processes that occur between soil, vegetation, and surface water near ground level. They measure the transfer of heat, gas, and liquid between these surfaces. Microscale meteorology often involves the study of chemistry.

Tracking air pollutants is an example of microscale meteorology. MIRAGE-Mexico is collaboration between meteorologists in the United States and Mexico. The program studies the chemical and physical transformations of gases and aerosols in the pollution surrounding Mexico City. MIRAGE-Mexico uses observations from ground stations, aircraft, and satellites to track pollutants.

Mesoscale Meteorology

Mesoscale phenomena range in size from a few kilometers to roughly 1,000 kilometers (620 miles). Two important phenomena are mesoscale convective complexes (MCC) and mesoscale convective systems (MCS). Both are caused by convection, an important meteorological principle.

Convection is a process of circulation. Warmer, less-dense fluid rises, and colder, denser fluid sinks. The fluid that most meteorologists study is air. (Any substance that flows is considered a fluid.) Convection results in a transfer of energy, heat, and moisture-the basic building blocks of weather.

In both an MCC and MCS, a large area of air and moisture is warmed during the middle of the day-when the sun angle is at its highest. As this warm air mass rises into the colder atmosphere, it condenses into clouds, turning water vapor into precipitation.

An MCC is a single system of clouds that can reach the size of the state of Ohio and produce heavy rainfall and flooding. An MCS is a smaller cluster of thunderstorms that lasts for several hours. Both react to unique transfers of energy, heat, and moisture caused by convection.

Synoptic Scale Meteorology

Synoptic-scale phenomena cover an area of several hundred or even thousands of kilometers. High- and low-pressure systems seen on local weather forecasts are synoptic in scale. Pressure, much like convection, is an important meteorological principle that is at the root of large-scale weather systems as diverse as hurricanes and bitter cold outbreaks.

Low-pressure systems occur where the atmospheric pressure at the surface of the Earth is less than its surrounding environment. Wind and moisture from areas with higher pressure seek low-pressure systems. This movement, in conjunction with the Coriolis force and friction, causes the system to rotate counter-clockwise in the Northern Hemisphere and clockwise in the Southern Hemisphere, creating a cyclone. Cyclones have a tendency for upward vertical motion. This allows moist air from the surrounding area to rise, expand and condense into water vapor, forming clouds. This movement of moisture and air causes the majority of our weather events.

Hurricanes are a result of low-pressure systems (cyclones) developing over tropical waters in the Western Hemisphere. The system sucks up massive amounts of warm moisture from the sea, causing convection to take place, which in turn causes wind speeds to increase and pressure to fall.

When these winds reach speeds over 119 kilometers per hour (74 miles per hour), the cyclone is classified as a hurricane.

Hurricanes can be one of the most devastating natural disasters in the Western Hemisphere. The National Hurricane Center, in Miami, Florida, regularly issues forecasts and reports on all tropical weather systems. During hurricane season, hurricane specialists issue forecasts and warnings for every tropical storm in the western tropical Atlantic and eastern tropical Pacific. Businesses and government officials from the United States, the Caribbean, Central America, and South America rely on forecasts from the National Hurricane Center.

High-pressure systems occur where the atmospheric pressure at the surface of the Earth is greater than its surrounding environment. This pressure has a tendency for downward vertical motion, allowing for dry air and clear skies.

Extremely cold temperatures are a result of high-pressure systems that develop over the Arctic and move over the Northern Hemisphere. Arctic air is very cold because it develops over ice and snow-covered ground. This cold air is so dense that it pushes against Earth's surface with extreme pressure, preventing any moisture or heat from staying within the system.

Meteorologists have identified many semi-permanent areas of high-pressure. The Azores high, for instance, is a relatively stable region of high pressure around the Azores, an archipelago in the mid-Atlantic Ocean. The Azores high is responsible for arid temperatures of the Mediterranean basin, as well as summer heat waves in Western Europe.

Global Scale Meteorology

Global scale phenomena are weather patterns related to the transport of heat, wind, and moisture from the tropics to the poles. An important pattern is global atmospheric circulation; the large-scale movement of air that helps distribute thermal energy (heat) across the surface of the Earth.

Global atmospheric circulation is the fairly constant movement of winds across the globe. Winds develop as air masses move from areas of high pressure to areas of low pressure. Global atmospheric circulation is largely driven by Hadley cells. Hadley cells are tropical and equatorial convection patterns. Convection drives warm air high in the atmosphere, while cool, dense air pushes lower in a constant loop. Each loop is a Hadley cell.

Hadley cells determine the flow of trade winds, which meteorologists forecast. Businesses, especially those exporting products across oceans, pay close attention to the strength of trade winds because they help ships travel faster. Westerlies are winds that blow from the west in the mid-latitudes. Closer to the Equator, trade winds blow from the northeast (north of the Equator) and the southeast (south of the Equator).

Meteorological Methods

Persistence Forecast

This is the simplest method to use and comes in quite useful; it starts with an assumption that present conditions will not change. Based on season expectations and averages, the weather tomorrow will be much the same as it is today. This works best in areas where the weather is not volatile or where it changes slowly. Southern California is a good example of this where conditions rarely change and seasonal alterations are less volatile and incremental with little variation noticeable on a day-to-day basis. Ideal for short-term forecasting, its limits are usually exposed when unusual weather fronts move in. It is not particularly useful for long-range forecasting.

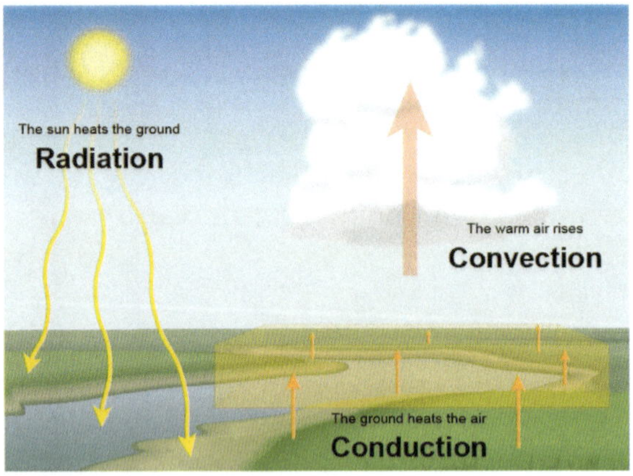

Trends Forecast

The trends forecasting method examines direction and speed of weather fronts, pressure bars, and cloud and precipitation build-up. This data is used to predict what the weather will be like in a certain area in a few hours or days based on what it is like now elsewhere. This relies on understanding of the conditions that cause a condition to intensify or dissipate as it progresses. This is best used to predict rain fronts. They will examine such elements as wind speed to predict when they will arrive. Weather is fairly predictable but can be subject to fluctuation based on the chaotic nature of new fronts forming and other forcings.

Numerical Weather Prediction

One of the most recent developments, it uses applied math to define weather conditions, patterns and trends. Today, meteorological organizations use computer modelling to on powerful computer systems to output a large number of predictions on all manner of atmospheric conditions. This hard data is then used to predict potential weather conditions short and long-term, and short and long-range. These supercomputers process thousands of computations every second to provide up-to-the-minute forecasts. They are not always right, but weather forecasts are right more often than not thanks to these computerized predictions. Often, errors are down to human error in inputting, insufficient data, and the chaotic nature of present weather conditions. When the equations are at fault, so will be the results. Other problems with the method include the lack of data from extreme environments. It's often difficult to acquire data from the middle of the ocean and tops of mountains, but satellite imagery can alleviate some of these problems.

Analog Method Forecasting

This is a method of comparison. In many ways, it is the opposite of Persistence Forecasting and works for some climate types more than others - particularly where erratic weather is common. Forecasters look to a date in the past (usually in the same season) to predict what the weather will be like tomorrow based on past experience. The assumption is that the change in weather pattern will mirror that of the past. If the weather is warm today but there is a change in wind direction or a cold front heading towards you, rather than assuming it will remain warm, forecasters will look for situations in the past where the same thing happened and attempt to predict how the weather might change. This can work well in predicting storms and other intense weather fronts. It has its problems, mostly because it relies on uniformity; if the weather has proven anything, it's that it is rarely uniform.

Climate-Based Methods

Our understanding of meteorological phenomena now has a new variable: climate change. We know that weather conditions are changing globally based on carbon emissions. It's understood that a warming climate will not lead to uniform warming everywhere. Some will become warmer and wetter, some warmer and drier. Some will experience slowing of their ocean jet streams and experience cooling - colder and wetter weather. Global climate change has the potential to change not just weather conditions, but change regional rules too. Erratic weather will become common and systems we have come to rely on to determine weather patterns may become near-useless. As the climate changes, meteorologists need to take seasonal averages over many years to understand what the weather might be like today or tomorrow or to predict long-term forecasts for future developments. This can also inform medical sciences and the spread of epidemics.

Uses of Meteorology

Weather Forecasting

The most obvious public face of the science of meteorology is in weather forecasting. Every time we turn on the news to understand how our local weather is going to look for today or next few days, we are utilizing one of the most common applications of meteorology. It applies many scientific methods and tools in attempting to predict, through atmospheric indicators, how the weather conditions will look one hour, one day or one week from now. The further into the future, the more erratic and less predictable the conditions become.

Meteorologists who work in this area collect quantitative data (mathematical information and statistics) on such information as air pressure, present weather, wind in the different levels of the atmosphere to create forecast models based on recognition of past patterns, mitigation of model bias. Weather people often get predictions wrong because of something called "Chaos Theory" (the concept that conditions can change wildly based on minor fluctuations in conditions). Newtonian physics once determined that systems were stable, but Einstein determined they are erratic, unpredictable (to a degree) and subject to external influences based on minute changes. Multiple models are used today to increase accuracy and super-fast computational processes can highlight up-to-the-minute changes.

We rely on weather warnings - for tornados, hurricanes, flooding, heavy rain and snow etc. for safety advice, protecting our lives and homes too.

Commodity Trading

Perhaps one of the most surprising ways in which meteorology is applied is in commodities trading. Stocks and shares trade is one area of employment for meteorologists, especially when the dealing with commodity crops such as coffee (affected by adverse weather conditions) and fuel (we use more in unusually cold winters). These organizations trade based on longer-term weather forecasting and what a crop harvest is going to look like in a certain year. Thales of Miletus was perhaps the first person to do this when he predicted a bumper crop for olives, bought a press, and made a lot of money in the process. It's an inexact science, because weather conditions that will create a bumper harvest for one crop may prove destructive for another. There are many examples of how meteorology has presented speculators with opportunities to make money.

Even smaller businesses such as clothing retailers and restaurants are utilizing data from meteorological data specialists. Targeted advertising is set to go out at certain times such as commercials for wet weather clothing during unusually wet weather and sun cream during unusually warm weather, not just based on typical seasonal trends.

Aviation Meteorology

This division of meteorology deals with military and commercial flying and weather conditions in the upper levels of the atmosphere. Even when the weather is good at ground level, it doesn't mean the same conditions apply 30,000ft. Aviation meteorology is the applied science that dictates air traffic - whether a route is safe or dangerous, at what times, and whether flights can be made at all. They will disseminate data about head and tail-winds, temperature changes, ice build-up (which can damage aircraft performance) and variation on the ground, air pressure variation across the world and through the atmosphere, visibility and local conditions advisory systems for pilots. They can dictate when it's unsafe to take off or unsafe to land (and find alternate airports such as the eruption of the Icelandic volcano in that causes havoc across North America and Europe.

Agricultural Meteorology

Few industries and areas of our lives are as dependent on changes in weather conditions as agriculture. Crops for food and for clothing are necessary to live and for business, providing livelihoods for those who grow crops - not just food, but also commodity crops such as cotton and coffee. Meteorology determines when farmers should sow, when they should reap, and what steps they will need to take to protect crops from erratic weather. They may need to engage in flood mitigation or effective water management during drought to protect from crop failure.

Throughout the season from sowing to harvest, farmers and agricultural workers must engage in proper crop management and monitoring. This means effective watering or drainage but also includes ensuring the right nutrients remain in the soil for the crop and for the season but is also based their forecast crop yields on weather conditions and how quickly they may respond to changing patterns. Neither does meteorology just apply to crop management; livestock management for milk production depends on weather conditions too. Finally, some use agricultural meteorology as an applied science to determine the relationships between a local environment, crops, soil types, soil profile, and understanding which crops can and cannot grow in certain types of soil.

Environmental Meteorology

Environmental meteorology is concerned with the study of pollution and its effects on the climate as a forcing of local, regional and national weather patterns. It will look at such aspects as variation in temperatures, water vapor density (humidity), speed and intensity of wind, and many other weather conditions and phenomena. It will also look at the physics of meteorological processes of acoustical, electrical, optical, and the thermodynamic processes of the atmosphere. Cloud formation, precipitation, weather conditions and much more. Rather than looking at just the weather conditions resulting from meteorology, it also examines the potential impacts of weather conditions on the environment and on climate. Extreme weather can change a landscape significantly and therefore alter the weather patterns. Long-term and large-scale modelling, data accumulation and analyses and modelling feature heavily in environmental modelling.

Hydrometeorology

This is the subdivision of meteorology that fuses with the science of hydrology, examining how water transfers from dry land (evaporation) and lower echelons of the atmosphere. This examines cloud formation and the processes that lead to water-based natural phenomena, but it also predicts projects and studies water hazards such as flooding and tropical cyclones, but also the effects of drought and land desertification - all these things concern precipitation and hydrology. Common applications behind this branch include "the water budget" - the account of all water transferring in or out of a water asset. Too much and it will flood, too little and drought ensues. Hydrometeorologists also monitor variation, quantity, intensity and distribution of rainfall. Examination of snowstorms also falls within the remit of hydrometeorology.

This is multidisciplinary, using applied math, statistics, but also computer data modeling. Problems with the cross-disciplinary approach and lack of broad expertise are being addressed by some projects such as DRIHM which aims to use big data and broad methodology to improve hydrometeorological forecasting for the future and mitigate the effects of extreme weather.

Synoptic Meteorology

The Synoptic Scale is the series of lines we see on a weather forecast map. Rather like contour lines on a map, how close or far apart they are, determines the weather pattern. In cartography, these contours denote the steepness of the elevation. In meteorology, they serve a similar purpose - denoting density. Synoptic Meteorology examines large-scale weather systems and their impact on larger areas. It looks at the systems that form to create hurricanes and cyclones, including weather fronts, their directions and jet streams. Conditions must be right from a variety of directions and sources to determine whether or not a certain weather system will develop. It examines the structures of the atmosphere and its behaviors in attempting to forecast systems. Weather forecasting needs to look at the conditions way beyond the area of study to understand what the weather may be like in a certain area. Synoptic Meteorology, then, takes a broader view.

Maritime/Marine Meteorology

Where conditions on land are vital for agricultural workers, those who work at sea - industrial fishing, oil and gas rigs, military navy, commercial shipping, all need up-to-date information on the weather conditions where they are or where they are going to dictate operations. They need to take action to avoid or mitigate the effects of extreme weather at sea, such as storms and hurricanes, cyclones, icebergs, the jet stream and much more. For commercial fishing, the problems can go beyond merely avoiding terrible storms in the name of safety. Business decisions must be made while taking on board changes in sea and freshwater weather systems. Even after these conditions have passed, fish stocks may be higher or lower than they were prior to the storm. Such information is also required for leisure boat activities and public transportation across water bodies. This information is not just useful for people who work at sea; many weather fronts develop on the ocean and then hit land, affecting activities on the shore.

Military Meteorology

Armed forces all around the world rely heavily on weather conditions and meteorological forecasts to plan military operations and training exercises. Some of the biggest upsets in military history have come from adverse weather conditions. In World War II, the Nazi Regime troops were held back by an advancing Russian winter at Stalingrad. A century earlier, Napoleon had the same problem. The Spanish Armada's proposed invasion of England in 1588 failed because of storms in the channel - so did Julius Caesar's invasion of Britannia. On the flipside, the D-Day landings in France during WWII were meticulously planned and executed after lengthy examination of the weather conditions in the English Channel. Originally set for June 5th 1944, commanders postponed due to weather warnings from meteorologists. At the time of the postponement, the weather was calm, but it did turn bad around the originally proposed time of mission launch. No matter how well-trained the forces, military commanders the world over do not underestimate the impact that weather conditions can have.

Nuclear Meteorology

How do we detect radioactive particles in the atmosphere? How do we determine their impact on the environment? Radiation is a natural phenomenon; this young subdivision of meteorology arose during the 19th and 20th centuries as humanity began to understand radiation waves, their sources and effects, and of course the results of nuclear testing since the 1930s. Nuclear meteorology investigates the distribution of radioactive gases and aerosols. They are present at nuclear energy production facilities, monitoring the reactors for radiology leaks and predicting their effects on the environment. They work to ensure environmental compliance at any facility where nuclear technology is in use. Most of all, they monitor pollution distribution by examining air currents and turbulence and predicting where the material will spread. They were integral to examining the spread of nuclear material following the meltdown of Chernobyl and relaying reports to European governments about the disaster.

Renewable Energy

As supplies of fossil fuel run out, the world can expect to move increasingly towards renewable energy. In 2018, this is a growth area but has a long way to go to catch up with conventional energy use. Most renewable energy sources are highly dependent on weather conditions. In some cases, that begins from even before the resources are planned and built. Wind farms need to be placed in areas used experiencing high winds. Solar farms need to go in areas that get a lot of sunshine. Hydroelectric power requires regular, sustainable and predictable water sources. Meteorologists examine local weather system history to determine sites of best placement. They must also take into account freak, erratic and variation in weather. Calm weather reduces the capacity of wind turbines, overcast, cold and wet weather reduce solar panel effectiveness, and drought will reduce the effectiveness of hydroelectric systems. Biofuel such as elephant grass needs the right climate and weather conditions for a sustainable crop and long-term productivity. In cold weather conditions when demand is high, there is potential for production to be low depending on the methods used. Finally, errors in forecasting can lead to reduced accessibility and lost money for producers. Therefore, renewable energy planning and development is and will remain fundamental to renewable energy at every stage.

Extreme Weather and Disaster Management

Extreme weather is a fact of life. California's on-going drought and forest fires continue to spread year on year. Hurricanes such as Katrina cause high winds and flood urban centers. In 2017, no country was untouched by natural disaster somewhere within its borders. Meteorology is useful in examining and planning the spread of extreme weather beforehand and offering advice before it happens, the science is just as important during and after. Disaster relief organizations such as FEMA need to understand weather conditions while they plan relief efforts. As with military meteorology above, the weather alone can be the difference between success and failure. When weather conditions are part of the plan with a disaster relief strategy, professionals are able to provide relief where they can and to do so safely without risking the lives of recipients or their own. Planning will often begin days in advance to see when the worst conditions have subsided and may be subject to change each day with fluctuating weather conditions.

2

Meteorological Phenomena

The observable weather events which are studied within meteorology are known as meteorological phenomena. A few of them are heat wave, drought, cyclone, acid rain, thunderstorm and thundersnow. The chapter closely examines these major meteorological phenomena to provide an extensive understanding of the subject.

Meteorological phenomenon are often categorized, examined, and forecasted according to their average horizontal size and life span.

Heat Waves

A heat wave is a period of unusually hot weather lasting two or more days. The temperature is hotter than average for the region. In an extreme heat wave, temperatures break records. The frequency, severity, and length of heat waves are increasing.

Heat and humidity already kill an average of 1,300 people in the U.S. each year. Deadly heat waves of at least 20 days affect 30% of the earth's population. That will increase to 75% by 2100 if global warming isn't stopped.

National Weather Service Warnings

The U.S. National Weather Service notifies residents of upcoming heat waves. It uses a Heat Index Value also called the apparent temperature. It factors in humidity with the air temperature in degrees Fahrenheit.

The NWS issues a Heat Advisory when the heat index reaches a certain level within the next 12 to 24 hours. The trigger point depends on average temperatures and is different for each locality. The advisory may be issued for lower temperatures if it is early in the season or during a multi-day heat wave. A heat advisory means that people could be affected by heat. It warns them to take precautions. It also triggers public safety regulations. These can include a ban on evictions and electricity shutoffs.

The NWS issues an Excessive Heat Warning when higher heat levels are expected within the next 12 to 24 hours. At those temperatures, some people can be seriously affected or die if they don't take precautions. The warning alerts hospitals to prepare for an increase in emergency calls. It activates programs that check on the home-bound. Some cities will open public air-conditioned centers. It also triggers the same public safety regulations as a heat advisory.

An Excessive Heat Watch is a warning issued one to two days in advance of the heat wave.

Causes of Heat Wave

A heat wave is caused by a high-pressure system that hovers over an area. It traps heat beneath it like an oven. High-pressure systems force air downward. Hot air on the ground cannot escape into higher levels. Without rising air, there is no rain or clouds. The sun just bakes the area until a new pressure system is strong enough to push the high-pressure system away.

Climate change increases heat waves by increasing the Earth's average temperature. How much has it warmed? Since the 1880s, the earth's average temperature has risen 2.1 F or 1 degree Celsius. Global warming is occurring at a faster rate than at any other time in the Earth's history.

In the northern hemisphere, global warming is changing the jet stream. That's a river of wind formed when cold Arctic air meets warm southern air. But the Arctic is warming faster than the south. The jet stream slows down and becomes wavy. That allows heat waves to linger.

Effects

Over the past 30 years, heat waves have killed more people than all other weather-related natural disasters combined. Heat waves kill in four ways:

1. Heat stress causes dehydration and loss of body salt. That throws off the chemistry of the body.
2. As the body tries to cool, it taxes the heart. That can lead to failure in people with heart conditions.
3. When the core body temperature rises beyond 104 F, organs fail. The gut leaks toxins into the body, creating a deadly inflammatory response called heat stroke.
4. People drown while trying to cool off in lakes and rivers.

Heat waves are most likely to affect people who work or live outdoors. The Bureau of Labor Statistics estimates that more than 15 million U.S. workers have jobs that require them to be outdoors. Of those, 7 million are in construction, 5 million are in logistics, and 1.3 million are farmers. In July 2018, a United States Postal Service worker died on the job when the temperature hit 117 F.

Also at risk are those without air conditioning. Worker productivity declines by 2% for every degree Celsius above normal room temperature.

The elderly, injured, and children are most threatened. People on certain medications that curb their ability to sweat are also at risk. Alcoholic consumption worsens the effect of heat waves.

Heat waves contributed to the record wildfire season in the American West. The heat dried out vegetation, creating tinder for fires. Scientists were surprised that the heat was enough to overcome soil that was still moist from a wetter than normal winter. Wildfires are driven more by the temperature and moisture content in the air than by the moisture content in the soil.

Heat waves could be the reason behind the 45 to 75% decline in worldwide insect populations. Warmer temperatures make males less fertile by reducing sperm count, according to research by the University of East Anglia. It only took a 9 to 12 degree F spike over a five-day period to cut sperm count in half. A second spike almost sterilized the insects.

Drought

Meteorological drought is defined usually on the basis of the degree of dryness (in comparison to some "normal" or average amount) and the duration of the dry period. Definitions of meteorological drought must be considered as region specific since the atmospheric conditions that result in deficiencies of precipitation are highly variable from region to region.

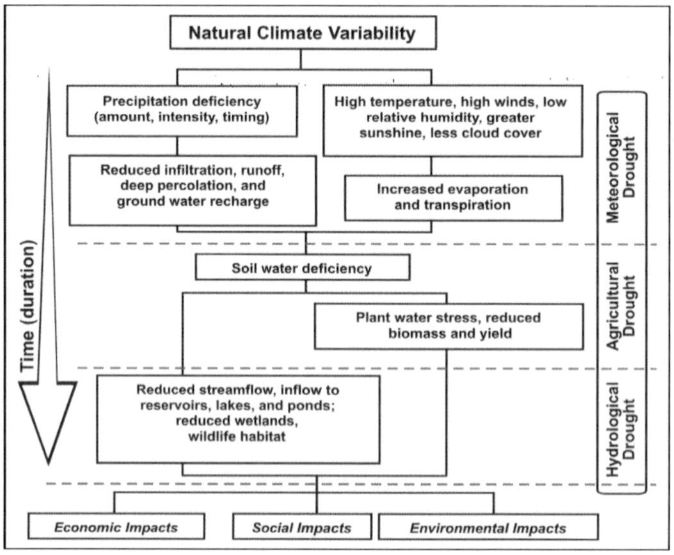

Sequence of drought occurrence and impacts for commonly accepted drought types. All droughts originate from a deficiency of precipitation or meteorological drought but other types of drought and impacts cascade from this deficiency.

For example, some definitions of meteorological drought identify periods of drought on the basis of the number of days with precipitation less than some specified threshold. This measure is only appropriate for regions characterized by a year-round precipitation regime such as a tropical rainforest, humid subtropical climate, or humid mid-latitude climate. Locations such as Manaus, Brazil; New Orleans, Louisiana (U.S.A.); and London, England, are examples.

Implications

Drought is a normal, recurring feature of the climate in most parts of the world. Having adequate drought mitigation strategies in place can greatly reduce the impact. Recurring or long-term drought can bring about desertification. Recurring droughts in the Horn of Africa have created grave ecological catastrophes, prompting massive food shortages, still recurring. To the north-west of the Horn, the Darfur conflict in neighbouring Sudan, also affecting Chad, was fuelled by decades of drought; combination of drought, desertification and overpopulation are among the causes of the Darfur conflict, because the Arab Baggara nomads searching for water have to take their livestock further south, to land mainly occupied by non-Arab farming peoples.

According to a UN climate report, the Himalayan glaciers that are the sources of Asia's biggest rivers—Ganges, Indus, Brahmaputra, Yangtze, Mekong, Salween and Yellow—could disappear by 2035 as temperatures rise. Approximately 2.4 billion people live in the drainage basin of the Himalayan Rivers. India, China, Pakistan, Bangladesh, Nepal and Myanmar could experience floods followed by droughts in coming decades. Drought in India affecting the Ganges is of particular concern, as it provides drinking water and agricultural irrigation for more than 500 million people. Paradoxically, some proposed short-term solutions to global warming also carry with them increased chances of drought.

In 2005, parts of the Amazon basin experienced the worst drought in 100 years. A July 2006 article reported Woods Hole Research Center results showing that the forest in its present form could survive only three years of drought. Scientists at the Brazilian National Institute of Amazonian Research argue in the article that this drought response, coupled with the effects of deforestation on regional climate, are pushing the rainforest towards a "tipping point" where it would irreversibly start to die. It concludes that the rainforest is on the brink of being turned into savanna or desert, with catastrophic consequences for the world's climate. According to the WWF, the combination of climate change and deforestation increases the drying effect of dead trees that fuels forests fires.

Causes of Drought

Generally, rainfall is related to the amount of water vapor in the atmosphere, combined with the upward forcing of the air mass containing that water vapor. If either of these is reduced, the result is drought.

Factors include:

- Above-average prevalence of high pressure systems;
- Winds carrying continental, rather than oceanic air masses (i.e. reduced water content);
- El Nino (and other oceanic temperature cycles);
- Deforestation.

Stages of Drought

Ship stranded by the retreat of the Aral Sea.

As a drought persists, the conditions surrounding it gradually worsen and its impact on the local population gradually increases. Droughts go through three stages before their ultimate cessation:

- Meteorological drought is brought about when there is a prolonged period with less than average precipitation. Meteorological drought usually precedes the other kinds of drought.
- Agricultural droughts are droughts that affect crop production or the ecology of the range.

This condition can also arise independently from any change in precipitation levels when soil conditions and erosion triggered by poorly planned agricultural endeavours cause a shortfall in water available to the crops. However, in a traditional drought, it is caused by an extended period of below average precipitation.

- Hydrological drought is brought about when the water reserves available in sources such as aquifers, lakes and reservoirs falls below the statistical average. Like an agricultural drought, this can be triggered by more than just a loss of rainfall. For instance, Kazakhstan was recently awarded a large amount of money by the World Bank to restore water that had been diverted to other nations from the Aral Sea under Soviet rule. Similar circumstances also place their largest lake, Balkhash, at risk of completely drying out.

Consequences of Drought

Periods of drought can have significant environmental, economic and social consequences. The most common consequences include:

- Death of livestock;
- Reduced crop yields;
- Wildfires, such as Australian bushfires, are more common during times of drought;
- Shortages of water for industrial users;
- Desertification;
- Dust storms, when drought hits an area suffering from desertification and erosion;
- Malnutrition, dehydration and related diseases;
- Famine due to lack of water for irrigation;
- Social unrest;
- Mass migration, resulting in internal displacement and international refugees;
- War over natural resources, including water and food;
- Reduced electricity production due to insufficient available coolant;
- Snakes have been known to emerge and snakebites become more common.

The effect varies according to vulnerability. For example, subsistence farmers are more likely to migrate during drought because they do not have alternative food sources. Areas with populations that depend on subsistence farming as a major food source are more vulnerable to drought-triggered famine. Drought is rarely if ever the sole cause of famine; socio-political factors such as extreme widespread poverty play a major role. Drought can also reduce water quality, because lower water flows reduce dilution of pollutants and increase contamination of remaining water sources.

Drought Mitigation Strategies

- Desalination of sea water for irrigation or consumption.
- Drought monitoring: Continuous observation of rainfall levels and comparisons with current usage levels can help prevent man-made drought. For instance, analysis of water usage in Yemen has revealed that their water table (underground water level) is put at grave risk by over-use to fertilize their Khat crop. Careful monitoring of moisture levels can also help predict increased risk for wildfires, using such metrics as the Keetch-Byram Drought Index or Palmer Drought Index.
- Land use: Carefully planned crop rotation can help to minimize erosion and allow farmers to plant less water-dependent crops in drier years.
- Rainwater harvesting: Collection and storage of rainwater from roofs or other suitable catchments.
- Recycled water: Former wastewater (sewage) that has been treated and purified for reuse.
- Transvasement: Building canals or redirecting rivers as massive attempts at irrigation in drought-prone areas.
- Water restrictions: Water use may be regulated (particularly outdoors). This may involve regulating the use of sprinklers, hoses or buckets on outdoor plants, the washing of motor vehicles or other outdoor hard surfaces (including roofs and paths), topping up of swimming pools, and also the fitting of water conservation devices inside the home (including shower heads, taps and dual flush toilets).
- Cloud seeding: an artificial technique to induce rainfall.

Cyclone

Tropical Cyclone

Tropical cyclone, also called typhoon or hurricane is an intense circular storm that originates over warm tropical oceans and is characterized by low atmospheric pressure, high winds, and heavy rain. Drawing energy from the sea surface and maintaining its strength as long as it remains over warm water, a tropical cyclone generates winds that exceed 119 km (74 miles) per hour. In extreme cases winds may exceed 240 km (150 miles) per hour, and gusts may surpass 320 km (200 miles) per hour. Accompanying these strong winds are torrential rains and a devastating phenomenon known as the storm surge, an elevation of the sea surface that can reach 6 metres (20 feet) above normal levels. Such a combination of high winds and water makes cyclones a serious hazard for coastal areas in tropical and subtropical areas of the world. Every year during the late summer months (July–September in the Northern Hemisphere and January–March in the Southern Hemisphere), cyclones strike regions as far apart as the Gulf Coast of North America, north-western Australia, and eastern India and Bangladesh.

Wrecked houseboats and bent palm trees in Key West, Florida, show the effects of Hurricane Georges, September 25, 1998.

Tropical cyclones are known by various names in different parts of the world. In the North Atlantic Ocean and the eastern North Pacific they are called hurricanes, and in the western North Pacific around the Philippines, Japan, and China the storms are referred to as typhoons. In the western South Pacific and Indian Ocean they are variously referred to as severe tropical cyclones, tropical cyclones, or simply cyclones. All these different names refer to the same type of storm.

Port-Vila, Vanuatu: Cyclone Pam. A boy and his father amid the debris of their home that was destroyed by Cyclone Pam in Port-Vila, Vanuatu, March 2015.

Anatomy of a Cyclone

Typhoon Odessa in the western North Pacific Ocean, photographed from the U.S. space shuttle Discovery, August 30, 1985.

Tropical cyclones are compact, circular storms, generally some 320 km (200 miles) in diameter, whose winds swirl around a central region of low atmospheric pressure. The winds are driven by this low-pressure core and by the rotation of Earth, which deflects the path of the wind through a phenomenon known as the Coriolis force. As a result, tropical cyclones rotate in a counterclockwise (or cyclonic) direction in the Northern Hemisphere and in a clockwise (or anticyclonic) direction in the Southern Hemisphere.

The wind field of a tropical cyclone may be divided into three regions. First is a ring-shaped outer region, typically having an outer radius of about 160 km (100 miles) and an inner radius of about 30 to 50 km (20 to 30 miles). In this region the winds increase uniformly in speed toward the centre. Wind speeds attain their maximum value at the second region, the eyewall, which is typically 15 to 30 km (10 to 20 miles) from the centre of the storm. The eyewall in turn surrounds the interior region, called the eye, where wind speeds decrease rapidly and the air is often calm. These main structural regions.

Eye of the Cyclone

A characteristic feature of tropical cyclones is the eye, a central region of clear skies, warm temperatures, and low atmospheric pressure. Typically, atmospheric pressure at the surface of Earth is about 1,000 millibars. At the centre of a tropical cyclone, however, it is typically around 960 millibars, and in a very intense "super typhoon" of the western Pacific it may be as low as 880 millibars. In addition to low pressure at the centre, there is also a rapid variation of pressure across the storm, with most of the variation occurring near the centre. This rapid variation results in a large pressure gradient force, which is responsible for the strong winds present in the eyewall.

Horizontal winds within the eye, on the other hand, are light. In addition, there is a weak sinking motion, or subsidence, as air is pulled into the eyewall at the surface. As the air subsides, it compresses slightly and warms, so that temperatures at the centre of a tropical cyclone are some 5.5 °C (10 °F) higher than in other regions of the storm. Because warmer air can hold more moisture before condensation occurs, the eye of the cyclone is generally free of clouds. Reports of the air inside the eye being "oppressive" or "sultry" are most likely a psychological response to the rapid change from high winds and rain in the eyewall to calm conditions in the eye.

Eyewall

The most dangerous and destructive part of a tropical cyclone is the eyewall. Here winds are strongest, rainfall is heaviest, and deep convective clouds rise from close to Earth's surface to a height of 15,000 metres (49,000 feet). As noted above, the high winds are driven by rapid changes in atmospheric pressure near the eye, which creates a large pressure gradient force. Winds actually reach their greatest speed at an altitude of about 300 metres (1,000 feet) above the surface. Closer to the surface they are slowed by friction, and higher than 300 metres they are weakened by a slackening of the horizontal pressure gradient force. This slackening is related to the temperature structure of the storm. Air is warmer in the core of a tropical cyclone, and this higher temperature causes atmospheric pressure in the centre to decrease at a slower rate with height than occurs in the surrounding atmosphere. The lessened contrast in atmospheric pressure with altitude causes the horizontal pressure gradient to weaken with height, which in turn results in a decrease in wind speed.

Friction at the surface, in addition to lowering wind speeds, causes the wind to turn inward toward the area of lowest pressure. Air flowing into the low-pressure eye cools by expansion and in turn extracts heat and water vapour from the sea surface. Areas of maximum heating have the strongest updrafts, and the eyewall exhibits the greatest vertical wind speeds in the storm—up to 5 to 10 metres (16.5 to 33 feet) per second, or 18 to 36 km (11 to 22 miles) per hour. While such velocities are much less than those of the horizontal winds, updrafts are vital to the existence of the towering convective clouds embedded in the eyewall. Much of the heavy rainfall associated with tropical cyclones comes from these clouds.

The upward movement of air in the eyewall also causes the eye to be wider aloft than at the surface. As the air spirals upward it conserves its angular momentum, which depends on the distance from the centre of the cyclone and on the wind speed around the centre. Since the wind speed decreases with height, the air must move farther from the centre of the storm as it rises.

When updrafts reach the stable tropopause (the upper boundary of the troposphere, some 16 km [10 miles] above the surface), the air flows outward. The Coriolis force deflects this outward flow, creating a broad anticyclonic circulation aloft. Therefore, horizontal circulation in the upper levels of a tropical cyclone is opposite to that near the surface.

Rainbands

In addition to deep convective cells (compact regions of vertical air movement) surrounding the eye, there are often secondary cells arranged in bands around the centre. These bands, commonly called rainbands, spiral into the centre of the storm. In some cases the rainbands are stationary relative to the centre of the moving storm, and in other cases they seem to rotate around the centre. The rotating cloud bands often are associated with an apparent wobbling of the storm track. If this happens as the tropical cyclone approaches a coastline, there may be large differences between the forecast landfall positions and actual landfall.

As a tropical cyclone makes landfall, surface friction increases, which in turn increases the convergence of airflow into the eyewall and the vertical motion of air occurring there. The increased convergence and rising of moisture-laden air is responsible for the torrential rains associated with tropical cyclones, which may be in excess of 250 mm (10 inches) in a 24-hour period. At times a storm may stall, allowing heavy rains to persist over an area for several days. In extreme cases, rainfall totals of 760 mm (30 inches) in a five-day period have been reported.

Life of a Cyclone

A circulation system goes through a sequence of stages as it intensifies into a mature tropical cyclone. The storm begins as a tropical disturbance, which typically occurs when loosely organized cumulonimbus clouds in an easterly wave begin to show signs of a weak circulation. Once the wind speed increases to 36 km (23 miles) per hour, the storm is classified as a tropical depression. If the circulation continues to intensify and the wind speeds exceed 63 km (39 miles) per hour, then the system is called a tropical storm. Once the maximum wind speed exceeds 119 km (74 miles) per hour, the storm is classified as a tropical cyclone.

There are six conditions favourable for this process to take place. The conditions are listed first below, and then their dynamics are described in greater detail:

1. The temperature of the surface layer of ocean water must be 26.5 °C (80 °F) or warmer, and this warm layer must be at least 50 metres (150 feet) deep.

2. A pre-existing atmospheric circulation must be located near the surface warm layer.
3. The atmosphere must cool quickly enough with height to support the formation of deep convective clouds.
4. The middle atmosphere must be relatively humid at a height of about 5,000 metres (16,000 feet) above the surface.
5. The developing system must be at least 500 km (300 miles) away from the Equator.
6. The wind speed must change slowly with height through the troposphere—no more than 10 metres (33 feet) per second between the surface and an altitude of about 10,000 metres (33,000 feet).

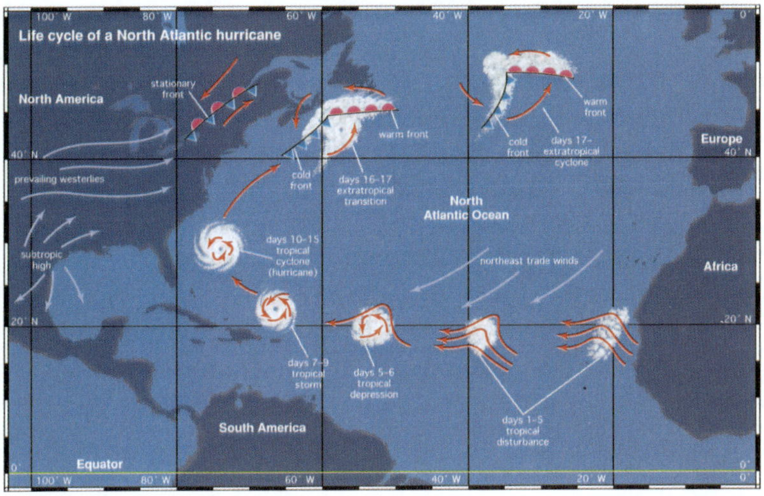

Formation

The fuel for a tropical cyclone is provided by a transfer of water vapour and heat from the warm ocean to the overlying air, primarily by evaporation from the sea surface. As the warm, moist air rises, it expands and cools, quickly becoming saturated and releasing latent heat through the condensation of water vapour. The column of air in the core of the developing disturbance is warmed and moistened by this process. The temperature difference between the warm, rising air and the cooler environment causes the rising air to become buoyant, further enhancing its upward movement.

If the sea surface is too cool, there will not be enough heat available, and the evaporation rates will be too low to provide the tropical cyclone enough fuel. Energy supplies will also be cut off if the warm surface water layer is not deep enough, because the developing tropical system will modify the underlying ocean. Rain falling from the deep convective clouds will cool the sea surface, and the strong winds in the centre of the storm will create turbulence. If the resulting mixing brings cool water from below the surface layer to the surface, the fuel supply for the tropical system will be removed.

The vertical motion of warm air is by itself inadequate to initiate the formation of a tropical system. However, if the warm, moist air flows into a pre-existing atmospheric disturbance, further development will occur. As the rising air warms the core of the disturbance by both release of latent heat and direct heat transfer from the sea surface, the atmospheric pressure in the centre of the disturbance becomes lower. The decreasing pressure causes the surface winds to increase, which in turn increases the vapour and heat transfer and contributes to further rising of air. The warming of the core and the increased surface winds thus reinforce each other in a positive feedback mechanism.

Intensification

The dynamics of a tropical cyclone rely on the exterior of a storm being cooler than its core, so it is necessary that the temperature of the atmosphere drop sufficiently rapidly with height. The warm, saturated air rising in the centre of the circulation tends to keep rising as long as the surrounding air is cooler and heavier. This vertical movement allows deep convective clouds to develop. The rising air in the core also draws in some air from the surrounding atmosphere at altitudes of around 5,000 metres (16,000 feet). If this external air is relatively humid, the circulation will continue to intensify. If it is sufficiently dry, then it may evaporate some of the water drops in the rising column, causing the air to become cooler than the surrounding air. This cooling will result in the formation of strong downdrafts that will disrupt the rising motion and inhibit development.

For the development of the rapid rotation characteristic of tropical cyclones, the low-pressure centre must be located at least 500 km (300 miles) away from the Equator. If the initial disturbance is too close to the Equator, then the effect of the Coriolis force will be too small to provide the necessary spin. The Coriolis force deflects the air that is being drawn into the surface low-pressure centre, setting up a cyclonic rotation. In the Northern Hemisphere the direction of the resulting circulation around the low is counter clockwise, and in the Southern Hemisphere it is clockwise.

A final requirement for the intensification of tropical cyclones is that there must be little change in the wind speed with height above the surface. If the winds increase too much with altitude, the core of the system will no longer be vertically aligned over the warm surface that provides its energy. The area being warmed and the surface low-pressure centre will move apart, and the positive feedback mechanism described above will be suppressed. Conditions in the tropics that encourage the development of tropical cyclones include a typically minor north-to-south variation in temperature. This relative lack of a temperature gradient causes wind speed to remain relatively constant with height.

Dissipation

Tropical cyclones dissipate when they can no longer extract sufficient energy from warm ocean water. As mentioned above, a tropical cyclone can contribute to its own demise by stirring up deeper, cooler ocean waters. In addition, a storm that moves over land will abruptly lose its fuel source and quickly lose intensity.

A tropical cyclone that remains over the ocean and moves into higher latitudes will change its structure and become extra tropical as it encounters cooler water. The transformations from a tropical to an extra tropical cyclone is marked by an increase in the storm's diameter and by a change in shape from circular to comma- or v-shaped as its rain bands reorganize. An extra tropical cyclone typically has a higher central pressure and consequently has lower wind speeds. Extra tropical cyclones, which are fuelled by a north-to-south variation of temperature, weaken and dissipate in a few days.

Tropical Cyclone Damage

Horizontal Wind

High winds cause some of the most dramatic and damaging effects associated with tropical cyclones. In the most intense tropical cyclones, sustained winds may be as high as 240 km (150 miles) per hour, and gusts can exceed 320 km (200 miles) per hour. The length of time that a given location is exposed to extreme winds depends on the size of the storm and the speed at which it is moving. During a direct hit from a tropical cyclone, an area may endure high winds for several hours. In that time even the most solidly constructed buildings may begin to suffer damage. The force of the wind increases rapidly with its speed. Sustained winds of 100 km (62 miles) per hour exert a pressure of 718 pascals (15 pounds per square foot), while an approximate doubling of wind speed to 200 km (124 miles) per hour increases the pressure almost fivefold to 3,734 pascals. A building with a large surface area facing the wind may be subjected to immense forces. Some of the local variability in damage that is often observed during tropical cyclones is due to the direction that buildings face relative to the prevailing wind.

Hurricane Dennis: Strong winds from Hurricane Dennis, which struck Florida in June 2005.

Horizontal winds associated with a tropical cyclone vary in strength depending on the area of the storm in which they occur. The strongest winds are located in the right-forward quadrant of the storm, as measured along the line that the storm is moving. The intensification of winds in this quadrant is due to the additive effect of winds from the atmospheric flow in which the storm is embedded. For example, in a hurricane approaching the East Coast of the United States, the highest and most damaging winds are located to the northeast of the storm centre.

Tornadoes

The intense sustained winds present near the centre of tropical cyclones are responsible for inflicting heavy damage, but there is another wind hazard associated with these storms—tornadoes. Most tropical disturbances that reach storm intensity have tornadoes associated with them when they make landfall, though the tornadoes tend to be weaker than those observed in the Midwestern United States. The number of tornadoes varies, but about 75 percent of tropical cyclones generate fewer than 10. The largest number of tornadoes associated with a tropical cyclone was 141, reported in 1967 as Hurricane Beulah struck the Texas Gulf Coast in the United States.

Tornadoes can occur in any location near the centre of the storm. At distances greater than 50 km (30 miles) from the centre, they are confined to the northeast quadrant of Northern Hemisphere storms and to the southwest quadrant of Southern Hemisphere storms. How the tornadoes are generated is not clear, but surface friction probably plays a role by causing the wind to slow as the tropical cyclone makes landfall. Wind speeds near the surface decrease while those at higher levels are less affected, setting up a low-level horizontal rotation that becomes tilted into the vertical by updrafts, thus providing the concentrated spin required for a tornado.

Gusts, Downbursts and Swirls

In addition to tornadoes, tropical cyclones generate other localized damaging winds. When a tropical cyclone makes landfall, surface friction decreases wind speed but increases turbulence; this allows fast-moving air aloft to be transported down to the surface, thereby increasing the strength of wind gusts. There is also evidence of tropical cyclone downbursts, driven by evaporative cooling of air. These downbursts are similar to microbursts that may occur during severe thunderstorms. The winds associated with them typically flow in a different direction than those of the cyclone, allowing them to be identified. Other small-scale wind features associated with tropical cyclones are swirls. These are very small, intense, and short-lived vortices that occur under convective towers embedded in the eyewall. They are not classified as tornadoes because their peak winds last only a few seconds. Swirls may rotate in either a counter clockwise or a clockwise direction, and their peak winds are estimated to approach 320 km (200 miles) per hour.

Storm Surge

In coastal regions an elevation of sea level—the storm surge—is often the deadliest phenomenon associated with tropical cyclones. A storm surge accompanying an intense tropical cyclone can be as high as 6 metres (20 feet). Most of the surge is caused by friction between the strong winds in the storm's eyewall and the ocean surface, which piles water up in the direction that the wind is blowing. For tropical cyclones in the Northern Hemisphere this effect is largest in the right-forward quadrant of the storm because the winds are strongest there. In the Southern Hemisphere the left-forward quadrant has the largest storm surge.

A fertilizer plant at the port of Paradip, India, inundated by a storm surge after the Orissa cyclone of October 29–30, 1999.

A small part of the total storm surge is due to the change in atmospheric pressure across the tropical cyclone. The higher atmospheric pressure at the edges of the storm causes the ocean surface to bulge under the eye, where the pressure is lowest. However, the magnitude of this pressure-induced surge is minimal because the density of water is large compared with that of air. A pressure drop of 100 millibars across the diameter of the storm causes the sea surface under the eye to rise about 1 metre (3 feet).

Flooding caused by the storm surge is responsible for most of the deaths associated with tropical cyclone landfalls. Extreme examples of storm surge fatalities include 6,000 deaths in Galveston, Texas, in 1900 and the loss of more than 300,000 lives in East Pakistan (now Bangladesh) in 1970 from a storm surge that was estimated to be 9 metres (30 feet) high. Improvements in forecasting the expected height of storm surges and the issuing of warnings are necessary as the population of coastal areas continues to increase.

Rainfall

Tropical cyclones typically bring large amounts of water into the areas they affect. Much of the water is due to rainfall associated with the deep convective clouds of the eyewall and with the rainbands of the outer edges of the storm. Rainfall rates are typically on the order of several centimetres per hour with shorter bursts of much higher rates. It is not uncommon for totals of 500 to 1,000 mm (20 to 40 inches) of rain to be reported over some regions. Rainfall rates such as these may overwhelm the capacity of storm drains, resulting in local flooding. Flooding may be particularly severe in low-lying regions such as in Bangladesh and the Gulf Coast of the United States. It is also a problem in areas where mountains and canyons concentrate the rainfall, as occurred in 1998 when floods caused by rains from Hurricane Mitch washed away entire towns in Honduras.

Another source of high precipitation may be provided by the migration of moist air from the clouds of the mature tropical cyclone. When this moisture moves into areas of low pressure at higher latitudes, significant precipitation may result. An example of this occurred in 1983, when the remnants of the eastern Pacific Hurricane Octave moved into a Pacific cold front that had stalled over the south-western United States, drenching the Arizona desert with 200 mm (8 inches) of rain in a three-day period. On average, that region receives 280 mm (11 inches) of rain in an entire year.

Location and Patterns of Tropical Cyclones

Ocean Basins and Peak Seasons

Tropical oceans spawn approximately 80 tropical storms annually, and about two-thirds are severe (category 1 or higher on the Saffir-Simpson scale of intensity). Almost 90 percent of these storms form within 20° north or south of the Equator. Poleward of those latitudes, sea surface temperatures are too cool to allow tropical cyclones to form, and mature storms moving that far north or south will begin to dissipate. Only two tropical ocean basins do not support tropical cyclones, because they lack waters that are sufficiently warm. The Peru Current in the eastern South Pacific and the Benguela Current in the South Atlantic carry cool water Equatorward from higher latitudes and so deter tropical cyclone development. The Pacific Ocean generates the greatest number of tropical storms and cyclones. The most powerful storms, sometimes called super typhoons, occur in the western Pacific. The Indian Ocean is second in the total number of storms, and the Atlantic Ocean ranks third.

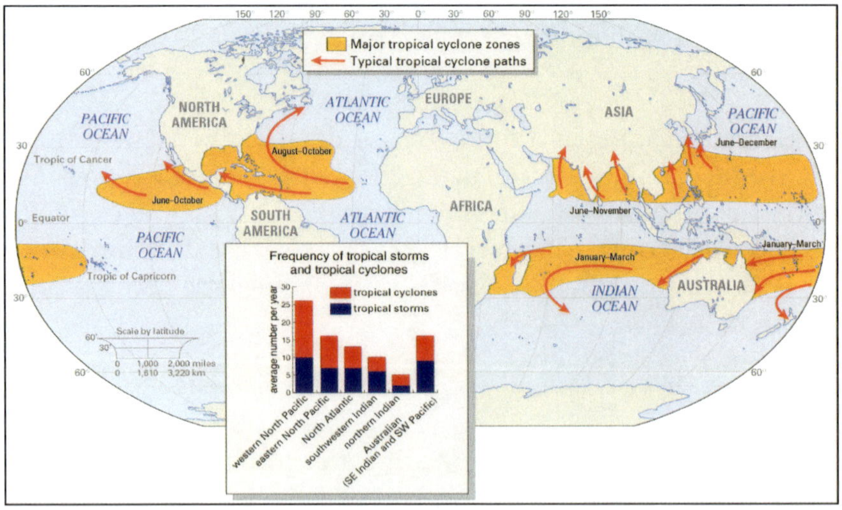

Major tracks and frequency of tropical cyclones (hurricanes and typhoons) and tropical storms.

Tropical cyclones are warm season phenomena. The peak frequency of these storms occurs after the maximum in solar radiation is received for the year, which occurs on June 22 in the Northern Hemisphere and December 22 in the Southern Hemisphere. The ocean surface reaches its maximum temperature several weeks after the solar radiation maximum, so most tropical cyclones occur during the late summer to early fall—that is, from July to September in the Northern Hemisphere and from January to March in the Southern Hemisphere.

Favourable Wind Systems

The lower latitudes are favourable for the generation of tropical cyclones not only because of their warm ocean waters but also because of the general atmospheric circulation of the region. Tropical cyclones originate from loosely organized, large-scale circulation systems such as those associated with the strong, low-level easterly jet over Africa. This jet generates easterly waves—regions of low atmospheric pressure that have a maximum intensity at an altitude of about 3,600 metres (12,000 feet) and a horizontal extent of about 2,400 km (1,500 miles). Most of the tropical cyclones in the Atlantic and eastern North Pacific begin as easterly waves. Given favourable conditions, an easterly wave may intensify and contract horizontally, ultimately resulting in the characteristic circulation of a tropical cyclone. In the western Pacific, large areas of upper-level low pressure help pull air from the centre of the developing disturbances and thus contribute to a drop in surface atmospheric pressure. It is these features, known as tropical upper tropospheric troughs, or TUTTs, that are responsible for the large number of tropical cyclones in the western Pacific.

In some cases, external geographic factors aid in development of tropical cyclones. The mountains of Mexico and Central America modify easterly waves that move through the Caribbean and into the eastern Pacific. This often results in closed circulations at low levels over the eastern Pacific Ocean, many of which develop into tropical cyclones.

Tropical Cyclone Tracks

Tropical cyclones in both the Northern and Southern Hemispheres tend to move westward and drift slowly poleward. Their motion is due in large part to the general circulation of Earth's atmosphere. Surface winds in the tropics, known as the trade winds, blow from east to west, and they are responsible for the general westward motion of tropical cyclones. For the poleward movement, two other factors are responsible. One is the presence of large-scale regions of subsiding air, known as subtropical highs, over the oceans poleward of the trade winds. These regions of high atmospheric pressure have anticyclonic circulations (that is, clockwise circulation in the Northern Hemisphere and counter clockwise in the Southern), so that winds on the western edges of these large-scale circulations move toward the poles. The second factor is the Coriolis force, which becomes progressively stronger at higher latitudes. The diameter of a tropical cyclone is large enough for the Coriolis force to influence its poleward side more strongly, and hence the tropical cyclone is deflected toward the pole. Once a tropical cyclone moves poleward of the subtropical high, it begins to move eastward under the influence of the middle-latitude westerlies (which blow toward the east). When the motion of a tropical cyclone changes from westward to eastward, the tropical cyclone is said to recurve.

Tropical cyclones in the Northern Hemisphere can travel to higher latitudes than in the Southern Hemisphere because of the presence of warm clockwise oceanic currents such as the Kuroshio and the Gulf Stream. In the North Atlantic the warm waters of the Gulf Stream supply energy to hurricanes as they move along the east coast of the United States, allowing them to survive for a longer time. It is not uncommon for very intense tropical systems to make landfall as far north as Boston (42 °N). On the other hand, hurricanes do not make landfall on the west coast of the United States even though prevailing winds over the North Pacific Ocean move eastward toward land. Instead, they tend to weaken rapidly as they recurve because they are moving over cooler ocean waters.

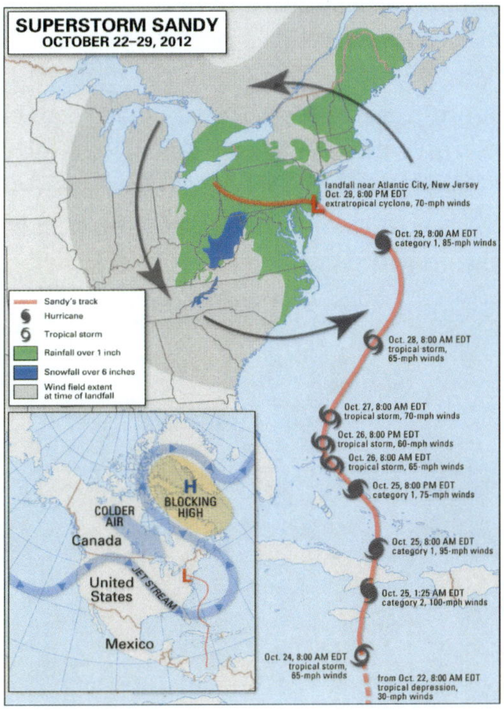

Superstorm Sandy: The evolution of Superstorm Sandy, from its origin as a tropical depression in the Caribbean Sea and its transformation into a hurricane.

Tracking and Forecasting

In the first half of the 20th century the identification of tropical cyclones was based on changes in weather conditions, the state of the sea surface, and reports from areas that had already been affected by the storm. This method left little time for advance warning and contributed to high death tolls. Observation networks and techniques improved with time; with the advent of weather satellites in the 1960s, the early detection and tracking of tropical cyclones was greatly improved.

Earth's Western Hemisphere: A view of Earth's Western Hemisphere from space showing Hurricane Linda approaching Baja California, Sept. 11, 1997.

Use of satellites and Aircraft

An array of geostationary satellites (those that remain over a fixed position on Earth) is operated by a number of countries. Each of these satellites provides continuous displays of Earth's surface in visible light and in infrared wavelengths. It is the latter that are most important in tracking the stages of tropical cyclone development. Infrared images show the temperatures of cloud tops, thus allowing the loosely organized convection associated with easterly waves to be detected by the presence of cold, high clouds. They also show the deep, organized convection characteristic of a tropical cyclone. Satellite images not only show a storm's location but also can be used to estimate its intensity because certain cloud patterns are characteristic of particular wind speeds.

"Supertyphoon" Chaba (right) approaching Japan and Typhoon Aere (left) hitting Taiwan, as photographed by the GOES-9 satellite, Aug. 25, 2004.

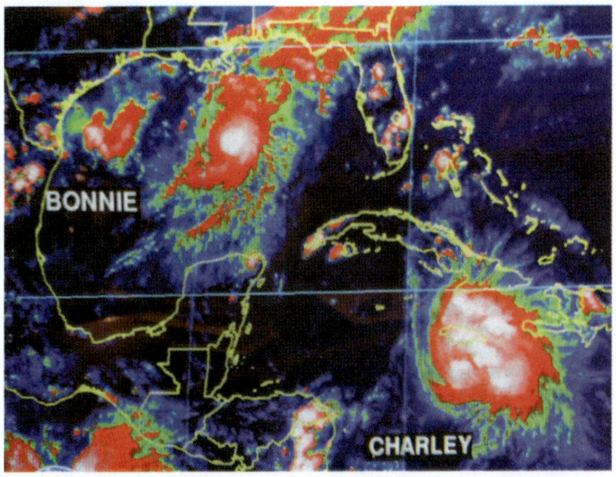

A screen at the U.S. National Hurricane Center, Miami, Florida, shows an infrared satellite image of Tropical Storm Bonnie (left) and Hurricane Charley (right), August 11, 2004.

Although satellite images provide general information on the location and intensity of tropical cyclones, detailed information on a storm's strength and structure must be obtained directly, using aircraft. This information is essential in providing the most accurate warnings possible. Operational reconnaissance is done only by the United States for storms that may affect its continental landmass. No other country does this type of reconnaissance. Tropical cyclones in other ocean basins occur over a larger region, and most countries do not have the financial resources to maintain research aircraft. When evidence of a developing circulation is detected in the Atlantic or Caribbean, a U.S. Air Force C-130 aircraft is dispatched to determine if a closed circulation is present. The centre of circulation is noted, and an instrument called a dropsonde is released through the bottom of the aircraft to measure the temperature, humidity, atmospheric pressure, and wind speed. In many cases, the naming of a tropical storm, or its upgrade from tropical storm to tropical cyclone, is based on aircraft observations.

Landfall Forecasts

Tropical storms developing in the world's ocean basins are tracked by various national weather services that have been designated Regional Specialized Meteorological Centres (RSMCs) by the World Meteorological Organization (WMO). The RSMCs are located at Miami, Florida, and Honolulu, Hawaii, U.S.; Tokyo, Japan; Nadi, Fiji; Darwin, Northern Territory, Australia; New Delhi, India; and Saint-Denis, Réunion. Warnings are also issued for more limited regions by Tropical Cyclone Warning Centres in a number of locations, including Port Moresby, Papua New Guinea; Wellington, New Zealand; and Perth, Western Australia, and Brisbane, Queensland, Australia. In addition, the Joint Typhoon Warning Centers in Hawaii are responsible for U.S. military forecasts in the western Pacific and Indian Oceans, which overlap a number of WMO regions of responsibility.

Forecasting hurricane landfall and providing warnings for storms that will affect the United States is done by the National Hurricane Center in Miami. Forecasters use a variety of observational information from satellites and aircraft to determine the current location and intensity of the storm. This information is used along with computer forecast models to predict the future path and intensity of the storm. There are three basic types of computer models. The simplest ones use statistical relations based on the typical paths of hurricanes in a region, along with the assumption that the current observed motion of the storm will persist. A second type of model, called a statistical-dynamical model, forecasts the large-scale circulation by solving equations that describe changes in atmospheric pressure, wind, and moisture. Statistical relations that predict the track of the storm based on the large-scale conditions are then used to forecast the storm's future position. A third type of model is a purely dynamic forecast model. In this model, equations are solved that describe changes in both the large-scale circulation and the tropical cyclone itself. Dynamic forecast models show the interaction of the tropical cyclone with its environment, but they require the use of large and powerful computers as well as very complete descriptions of the structure of the tropical cyclone and that of the surrounding environment. Computer models currently do well in forecasting the path of tropical cyclones, but they are not as reliable in forecasting changes in intensity more than 24 hours in advance.

Once forecasters have determined that a tropical cyclone is likely to make landfall, warnings are issued for the areas that may be affected. The forecasters provide a "best-track" forecast, which is an estimate of the track and maximum wind speed over a period of 72 hours based on all available observations and computer model results. Strike probability forecasts are issued that indicate probabilities (in percentages) that the tropical cyclone will affect a given area over a given time interval. These forecasts allow local authorities to begin warning and evacuation plans. As the storm approaches, a tropical cyclone watch is issued for areas that may be threatened. In especially vulnerable areas, evacuation may be initiated based on the watch. If tropical cyclone conditions are expected in an area within 24 hours, a tropical cyclone warning is issued. Once a warning is issued, evacuation is recommended for areas prone to storm surges and areas that may be isolated by high water.

Long-term Forecasts

Forecasts of expected numbers of Atlantic tropical cyclones are now being made well in advance of the start of each year's tropical cyclone season. The forecast model takes into account seasonal trends in factors related to tropical cyclone formation such as the presence of El Niño or La Niña oceanic conditions, amount of rainfall over Africa, winds in the lower stratosphere, and atmospheric pressure and wind tendencies over the Caribbean. Based on these factors, forecasts are issued concerning the expected numbers of tropical storms, tropical cyclones, and intense tropical cyclones for the Atlantic. These forecasts are issued in December, and they are revised in June and again in August of each year for the current Atlantic tropical cyclone season. The forecast model has displayed reasonable skill in predicting the total number of storms each season.

Climatic Variations and Tropical Cyclone Frequency

The number of tropical cyclones generated during a given a year has been observed to vary with certain climatic conditions that modify the general circulation of the atmosphere. One of these conditions is the intermittent occurrence of El Niño, an oceanic phenomenon characterized by the presence every few years of unusually warm water over the equatorial eastern Pacific. The presence of unusually cool surface waters in the region is known as La Niña. While the factors connecting El Niño and La Niña to tropical cyclones are complicated, there are a few general relationships. During years when El Niño conditions are present, upper-level winds over the Atlantic tend to be stronger than normal, which increases the vertical shear and decreases tropical cyclone activity. La Niña conditions result in weaker shear and enhanced tropical cyclone activity. The variation of sea surface temperature associated with El Niño and La Niña also changes the strength and location of the jet stream, which in turn alters the tracks of tropical cyclones. There are indications that El Niño and La Niña modulate tropical cyclone activity in other parts of the world as well. More tropical cyclones seem to occur in the eastern portion of the South Pacific during El Niño years, and fewer occur during La Niña years.

The possibility is being examined that changes in Earth's climate might alter the numbers, intensity, or paths of tropical cyclones worldwide. Increasing the amount of carbon dioxide and other greenhouse gases in the atmosphere through the burning of fossil fuels and other human activities may increase the global average temperature and the temperature of the sea surface. These potential changes would influence the maximum intensity reached by a tropical cyclone, which depends on both the sea surface temperature and the temperature of the upper troposphere. An increase in global temperature, however, could actually decrease the number of tropical cyclones, because any change in temperature would be accompanied by changes in Earth's general circulation. If tropical atmospheric circulation were to change in such a way as to increase the winds at upper levels, then there could be a decrease in tropical cyclone activity. An assessment by the World Meteorological Organization of the effect of climate change on tropical cyclones concluded that there is no evidence to suggest that an enhanced greenhouse effect will cause any major changes in the global location of tropical cyclone genesis or the total area of Earth's surface over which tropical cyclones form. Furthermore, while the maximum potential intensity of tropical cyclones may increase by 10 to 20 percent with a doubling of the concentration of carbon dioxide in the atmosphere, factors such as increased cooling due to ocean spray and changes in the vertical temperature variation may offset these effects.

Subtropical Cyclone

A subtropical cyclone is a weather system that has some characteristics of a tropical and an extratropical cyclone.

As early as the 1950s, meteorologists were uncertain whether they should be characterized as tropical or extratropical cyclones. They were officially recognized by the National Hurricane Center in 1972. Beginning in 2002, subtropical cyclones received names from the official tropical cyclone lists in the North Atlantic, South-west Indian Ocean, and South Atlantic basins.

Subtropical Storm Leslie in September 2018.

There are two definitions currently used for subtropical cyclones. Across the north Atlantic and southwest Indian Ocean, they require central convection fairly near the center and a warming core in the mid-levels of the troposphere. Across the eastern half of the northern Pacific, they require a mid-tropospheric cyclone to be cut off from the main belt of the westerlies and only a weak surface circulation. Subtropical cyclones have broad wind patterns with maximum sustained winds located farther from the center than typical tropical cyclones, and have no weather fronts linked into their center.

Since they form from initially extratropical cyclones which have colder temperatures aloft than normally found in the tropics, the sea surface temperatures required for their formation are lower than the tropical cyclone threshold by 3 °C (5 °F), lying around 23 °C (73 °F). This also means that subtropical cyclones are more likely to form outside the traditional bounds of the North Atlantic hurricane season. Subtropical cyclones are also observed to form in the South Atlantic; South Atlantic subtropical cyclones are observed in all months.

Formation of Subtropical Cyclone

Subtropical cyclones form in a wide band of latitude, mainly south of the 50th parallel in the northern hemisphere. Due to the increased frequency of cyclones which cut off from the main belt of the westerlies during the summer and fall, subtropical cyclones are significantly more frequent across the North Atlantic than the north western Pacific Ocean. In the eastern half of the North Pacific Ocean and north Indian Ocean, the older subtropical cyclone definition term is still used, which requires a weak circulation forming underneath a mid to upper-tropospheric low which has cut off from the main belt of the westerlies during the cold season (winter). In the case of the north Indian Ocean, the formation of this type of vortex leads to the onset of monsoon rains during the wet season. In the southern hemisphere, subtropical cyclones are regularly observed across southern portions of the Mozambique Channel.

Most subtropical cyclones form when a deep cold-core extratropical cyclone drops down into the subtropics. The system becomes blocked by a high latitude ridge, and eventually sheds its frontal boundaries as its source of cool and dry air from the high latitudes diverts away from the system. Temperature differences between the 500 hPa pressure level, or 5,900 meters (19,400 ft) above ground level, and the sea surface temperatures initially exceed the dry adiabatic lapse rate, which causes an initial round of thunderstorms to form at a distance east of the center. Due to the initial cold temperatures aloft, sea surface temperatures usually need to reach at least 20 °C (68 °F) for this initial round of thunderstorms. The initial thunderstorm activity moistens up the environment around the low, which destabilizes the atmosphere by reducing the lapse rate needed for convection. When the next shortwave or upper level jet streak (wind maximum within the jet stream) moves nearby, convection reignites closer to the center and the system develops into a true subtropical cyclone. The average sea surface temperature that helps lead to subtropical cyclogenesis is 24 °C (75 °F). If the thunderstorm activity becomes deep and persistent, allowing its initial low level warm core to deepen, tropical cyclogenesis is possible. The locus of formation for North Atlantic subtropical cyclones is out in the open ocean; the island of Bermuda is regularly impacted by these systems.

The South Atlantic environment for formation of subtropical cyclones has both stronger vertical wind shear and lower sea surface temperatures, yet subtropical cyclogenesis is regularly observed in the open ocean in the South Atlantic. A second mechanism for formation has been diagnosed for South Atlantic subtropical cyclones: lee cyclogenesis in the region of the Brazil Current.

Transition from Extratropical

By gaining tropical characteristics, an extratropical low may transit into a subtropical depression or storm. A subtropical depression/storm may further gain tropical characteristics to become a pure tropical depression or storm, which may eventually develop into a hurricane, and there are at least ten cases of tropical cyclones transforming into a subtropical cyclone (Tropical Storm Gilda in 1973, Subtropical Storm Four in 1974, Tropical Storm Jose in 1981, Hurricane Klaus in 1984, Tropical Storm Allison in 2001, Tropical Storm Lee in 2011, Hurricane Humberto in 2013, Tropical Storm Ian in 2016, Typhoon Jelawat in 2018, and Tropical Storm Gaemi in 2018). There have also been two recorded cases of a storm transitioning from tropical to extratropical to a subtropical cyclone; as seen with Hurricane Georges in 1980 and Hurricane Beryl in 2018. Generally, a tropical storm or tropical depression is not called subtropical while it is becoming extratropical, after hitting either land or colder waters. This transition normally requires significant instability through the atmosphere, with temperature differences between the underlying ocean and the mid-levels of the troposphere requiring over 20 °C, or 72 °F, of contrast in this roughly 5,900 meters (19,400 ft) layer of the lower atmosphere. The mode of the sea surface temperatures that subtropical cyclones form over is 23 °C (73 °F). Transition from subtropical cyclones into tropical cyclones occurs only in very rare cases over the South Atlantic Ocean, such as Hurricane Catarina in 2004.

Characteristics

Subtropical Storm Andrea peaking as a weak subtropical cyclone on May 20, 2019.

These storms can have maximum winds extending farther from the center than in a purely tropical cyclone and have no weather fronts linking directly to the center of circulation. In the Atlantic Basin, the United States NOAA classifies subtropical cyclones similarly to their tropical cousins, based on maximum sustained surface winds. Those with winds below 18 m/s, (65 km/h, 35 knots, or 39 mph) are called subtropical depressions, while those at or above this speed are referred to as subtropical storms.

Subtropical cyclones with hurricane-force winds of 33 m/s, (119 km/h, 64 knots, or 74 mph) or greater are not officially recognized by the National Hurricane Center. Once a subtropical storm intensifies enough to have hurricane-force winds, it is then automatically assumed to have become a fully tropical hurricane. Despite this, however, prior to the start of modern policies, two subtropical cyclones in the Atlantic hurricane database attained hurricane-force winds; a subtropical storm in 1968, and a subtropical storm in 1979.

Subtropical cyclones are also more likely than tropical cyclones to form outside of a region's designated hurricane season. Recent North Atlantic examples of this include the following storms:

- Subtropical Storm Ana (which became Tropical Storm Ana) in late-April of the 2003 hurricane season.
- Subtropical Storm Andrea in early-May of the 2007 hurricane season.
- Subtropical Storm Olga (which became Tropical Storm Olga) in mid-December of the 2007 hurricane season.
- Subtropical Storm Beryl (which became Tropical Storm Beryl) in late-May of the 2012 hurricane season.
- An unnamed subtropical storm in early-December of the 2013 hurricane season.
- Subtropical Storm Ana (which became Tropical Storm Ana) in early-May of the 2015 hurricane season.
- Subtropical Storm Alex (which became Hurricane Alex) in mid-January of the 2016 hurricane season.
- Subtropical Depression One (which became Tropical Storm Arlene) in mid-April of the 2017 hurricane season.
- Subtropical Storm Alberto (which became Tropical Storm Alberto) in late-May of the 2018 Atlantic hurricane season.
- Subtropical Storm Andrea in late-May of the 2019 Atlantic hurricane season.

Types of Subtropical Storm

Upper-level Low

The most common type of subtropical storm is an upper-level cold low with circulation extending to the surface layer and maximum sustained winds generally occurring at a radius of about 160 kilometers (99 mi) or more from the center. In comparison to tropical cyclones, such systems have a relatively broad zone of maximum winds that is located farther from the center, and typically have a less symmetric wind field and distribution of convection.

Mesoscale Low

A second type of subtropical cyclone is a mesoscale low originating in or near a frontolyzing zone of horizontal wind shear, also known as a dying frontal zone, with radius of maximum sustained winds generally less than 50 kilometers (31 mi). The entire circulation may initially have a diameter of less than 160 kilometers (99 mi). These generally short-lived systems may be either cold core or warm core, and in 1972 this type of subtropical cyclone was referred to as a "neutercane".

Kona Storm

A subtropical storm in December 2010, originally a Kona storm.

Kona storms (or Kona lows) are deep cyclones that form during the cool season of the central Pacific Ocean. A definition change in the term during the early 1970s makes categorization of the systems more complex, as many kona lows are extratropical cyclones, complete with their own weather fronts. Those across the northeast Pacific Ocean consider them subtropical cyclones as long as a weak surface circulation is present. Kona is a Hawaiian term for leeward, which explains the change in wind direction for the Hawaiian Islands from easterly to southerly when this type of cyclone is present.

Australian East Coast Lows

An Australian East Coast Low in June 2013.

Australian east coast lows (known locally as east coast lows and sometimes as east coast cyclones) are extratropical cyclones, the most intense of these systems have many of the characteristics of subtropical cyclones. They develop between 25°south and 40°south and within 5° of the Australian coastline, typically during the winter months. Each year there are about ten "significant impact" maritime lows. Explosive cyclogenesis is seen on average just once per year, but these storms cause significant wind and flood damage when they occur. Australian east coast cyclones vary in size from mesoscale (approximately 10 km to 100 km) to synoptic scale (approximately 100 km to 1,000 km). These storms which mostly affect the south east coast should not be confused with Australian region tropical cyclones which typically affect the northern half of the continent.

Extratropical Cyclone

Extratropical cyclones, sometimes called mid-latitude cyclones or wave cyclones, are low-pressure areas which, along with the anticyclones of high-pressure areas, drive the weather over much of the Earth. Extratropical cyclones are capable of producing anything from cloudiness and mild showers to heavy gales, thunderstorms, blizzards, and tornadoes. These types of cyclones are defined as large scale (synoptic) low pressure weather systems that occur in the middle latitudes of the Earth. In contrast with tropical cyclones, extratropical cyclones produce rapid changes in temperature and dew point along broad lines, called weather fronts, about the center of the cyclone.

A powerful extratropical cyclone over the North Pacific Ocean in January 2018, with an eye-like feature and a long cold front extending to the tropics.

Formation

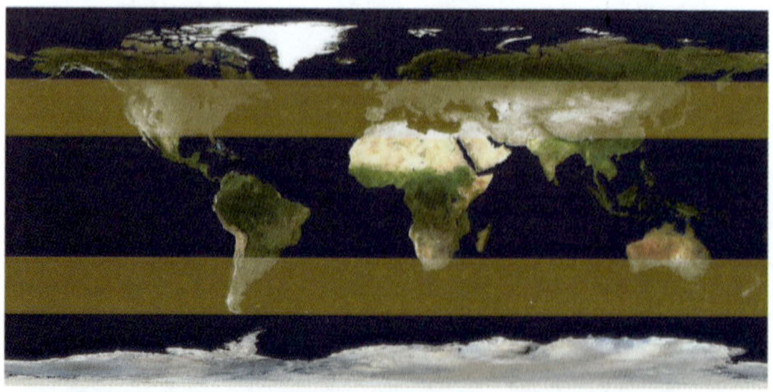

Approximate areas of extratropical cyclone formation worldwide.

Extratropical cyclones form anywhere within the extratropical regions of the Earth (usually between 30° and 60° latitude from the equator), either through cyclogenesis or extratropical transition. A study of extratropical cyclones in the Southern Hemisphere shows that between the 30th and 70th parallels, there are an average of 37 cyclones in existence during any 6-hour period. A separate study in the Northern Hemisphere suggests that approximately 234 significant extratropical cyclones form each winter.

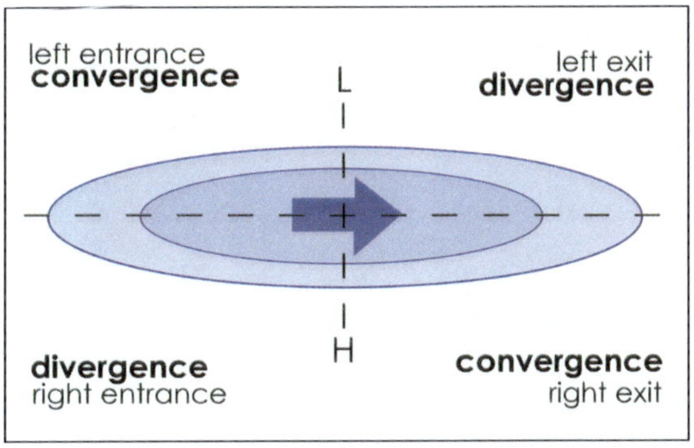

An upper level jet streak. DIV areas are regions of divergence aloft, which will lead to surface convergence and aid cyclogenesis.

Cyclogenesis

Extratropical cyclones form along linear bands of temperature/dewpoint gradient with significant vertical wind shear, and are thus classified as baroclinic cyclones. Initially, cyclogenesis, or low pressure formation, occurs along frontal zones near a favorable quadrant of a maximum in the upper level jetstream known as a jet streak. The favorable quadrants are usually at the right rear and left front quadrants, where divergence ensues. The divergence causes air to rush out from the top of the air column. As mass in the column is reduced, atmospheric pressure at surface level (the weight of the air column) is reduced. The lowered pressure strengthens the cyclone (a low pressure system). The lowered pressure acts to draw in air, creating convergence in the low-level wind field. Low-level convergence and upper-level divergence imply upward motion within the column, making cyclones tend to be cloudy. As the cyclone strengthens, the cold front sweeps towards the equator and moves around the back of the cyclone. Meanwhile, its associated warm front progresses more slowly, as the cooler air ahead of the system is denser, and therefore more difficult to dislodge. Later, the cyclones occlude as the poleward portion of the cold front overtakes a section of the warm front, forcing a tongue, or trowal, of warm air aloft. Eventually, the cyclone will become barotropically cold and begin to weaken.

Atmospheric pressure can fall very rapidly when there are strong upper level forces on the system. When pressures fall more than 1 millibar (0.030 inHg) per hour, the process is called explosive cyclogenesis, and the cyclone can be described as a (weather) bomb. These bombs rapidly drop in pressure to below 980 millibars (28.94 inHg) under favorable conditions such as near a natural temperature gradient like the Gulf Stream, or at a preferred quadrant of an upper level jet streak, where upper level divergence is best. The stronger the upper level divergence over the cyclone, the deeper the cyclone can become. Hurricane-force extratropical cyclones are most likely to form in the northern Atlantic and northern Pacific oceans in the months of December and January. On 14 and 15 December 1986, an extratropical cyclone near Iceland deepened to below 920 hectopascals (27 inHg), which is a pressure equivalent to a category 5 hurricane. In the Arctic, the average pressure for cyclones is 980 millibars (28.94 inHg) during the winter and 1,000 millibars (29.53 inHg) during the summer.

Extratropical Transition

Tropical cyclones often transform into extratropical cyclones at the end of their tropical existence, usually between 30° and 40° latitude, where there is sufficient forcing from upper-level troughs or shortwaves riding the Westerlies for the process of extratropical transition to begin. During this process, a cyclone in extratropical transition (known across the eastern North Pacific and North Atlantic oceans as the post-tropical stage), will invariably form or connect with nearby fronts and/or troughs consistent with a baroclinic system. Due to this, the size of the system will usually appear to increase, while the core weakens. However, after transition is complete, the storm may re-strengthen due to baroclinic energy, depending on the environmental conditions surrounding the system. The cyclone will also distort in shape, becoming less symmetric with time.

Hurricane Cristobal (2014) in the north Atlantic after completing its transition to an extratropical cyclone from a hurricane.

During extratropical transition, the cyclone begins to tilt back into the colder airmass with height, and the cyclone's primary energy source converts from the release of latent heat from condensation (from thunderstorms near the center) to baroclinic processes. The low pressure system eventually loses its warm core and becomes a cold-core system.

The peak time of subtropical cyclogenesis (the midpoint of this transition) in the North Atlantic is in the months of September and October, when the difference between the temperature of the air aloft and the sea surface temperature is the greatest, leading to the greatest potential for instability. On rare occasions, an extratropical cyclone can transit into a tropical cyclone if it reaches an area of ocean with warmer waters and an environment with less vertical wind shear. An example of this happening is in the 1991 Perfect Storm. The process known as "tropical transition" involves the usually slow development of an extratropically cold core vortex into a tropical cyclone.

The Joint Typhoon Warning Center uses the extratropical transition (XT) technique to subjectively estimate the intensity of tropical cyclones becoming extratropical based on visible and infrared satellite imagery. Loss of central convection in transitioning tropical cyclones can cause the Dvorak technique to fail; the loss of convection results in unrealistically low estimates using the Dvorak technique. The system combines aspects of the Dvorak technique, used for estimating tropical cyclone intensity, and the Hebert-Poteat technique, used for estimating subtropical cyclone intensity. The technique is applied when a tropical cyclone interacts with a frontal boundary or loses its central convection while maintaining its forward speed or accelerating. The XT scale corresponds to the Dvorak scale and is applied in the same way, except that "XT" is used instead of "T" to indicate that the system is undergoing extratropical transition. Also, the XT technique is only used once extratropical transition begins; the Dvorak technique is still used if the system begins dissipating without transition. Once the cyclone has completed transition and become cold-core, the technique is no longer used.

Structure

Surface Pressure and Wind Distribution

QuikSCAT image of typical extratropical cyclones over the ocean. Note the maximum winds are on the outside of the occlusion.

The windfield of an extratropical cyclone constricts with distance in relation to surface level pressure, with the lowest pressure being found near the center, and the highest winds typically just on the cold/poleward side of warm fronts, occlusions, and cold fronts, where the pressure gradient force is highest. The area poleward and west of the cold and warm fronts connected to extratropical cyclones is known as the cold sector, while the area equatorward and east of its associated cold and warm fronts is known as the warm sector.

Extratropical cyclones spin clockwise in the Southern Hemisphere, just like tropical cyclones.

The wind flow around an extratropical cyclone is counterclockwise in the northern hemisphere, and clockwise in the southern hemisphere, due to the Coriolis effect (this manner of rotation is generally referred to as cyclonic). Near this center, the pressure gradient force (from the pressure at the center of the cyclone compared to the pressure outside the cyclone) and the Coriolis force must be in an approximate balance for the cyclone to avoid collapsing in on itself as a result of the difference in pressure. The central pressure of the cyclone will lower with increasing maturity, while outside of the cyclone; the sea-level pressure is about average. In most extratropical cyclones, the part of the cold front ahead of the cyclone will develop into a warm front, giving the frontal zone (as drawn on surface weather maps) a wave-like shape. Due to their appearance on satellite images, extratropical cyclones can also be referred to as frontal waves early in their life cycle. In the United States, an old name for such a system is "warm wave".

In the northern hemisphere, once a cyclone occludes, a trough of warm air aloft—or "trowal" for short—will be caused by strong southerly winds on its eastern periphery rotating aloft around its northeast, and ultimately into its northwestern periphery (also known as the warm conveyor belt), forcing a surface trough to continue into the cold sector on a similar curve to the occluded front. The trowal creates the portion of an occluded cyclone known as its comma head, due to the comma-like shape of the mid-tropospheric cloudiness that accompanies the feature. It can also be the focus of locally heavy precipitation, with thunderstorms possible if the atmosphere along the trowal is unstable enough for convection.

Vertical Structure

Extratropical cyclones slant back into colder air masses and strengthen with height, sometimes exceeding 30,000 feet (approximately 9 km) in depth. Above the surface of the earth, the air temperature near the center of the cyclone is increasingly colder than the surrounding environment. These characteristics are the direct opposite of those found in their counterparts, tropical cyclones; thus, they are sometimes called "cold-core lows". Various charts can be examined to check the characteristics of a cold-core system with height, such as the 700 millibars (20.67 inHg) chart, which is at about 10,000 feet (3,048 meters) altitude.

Cyclone Evolution

A hurricane-force extratropical cyclone in January 2016 with a distinct eye-like feature, caused by a warm seclusion.

There are two models of cyclone development and lifecycles in common use—the Norwegian model and the Shapiro-Keyser Model.

Norwegian Cyclone Model

Of the two theories on extratropical cyclone structure and life cycle, the older is the Norwegian Cyclone Model, developed during World War I. In this theory, cyclones develop as they move up and along a frontal boundary, eventually occluding and reaching a barotropically cold environment. It was developed completely from surface-based weather observations, including descriptions of clouds found near frontal boundaries. This theory still retains merit, as it is a good description for extratropical cyclones over continental landmasses.

Shapiro-keyser Model

A second competing theory for extratropical cyclone development over the oceans is the Shapiro-Keyser model, developed in 1990. Its main differences with the Norwegian Cyclone Model are the fracture of the cold front, treating warm-type occlusions and warm fronts as the same, and allowing the cold front to progress through the warm sector perpendicular to the warm front. This model was based on oceanic cyclones and their frontal structure, as seen in surface observations and in previous projects which used aircraft to determine the vertical structure of fronts across the northwest Atlantic.

Warm Seclusion

A warm seclusion is the mature phase of the extratropical cyclone lifecycle. This was conceptualized after the ERICA field experiment of the late 1980s, which produced observations of intense marine cyclones that indicated an anomalously warm low-level thermal structure, secluded (or surrounded) by a bent-back warm front and a coincident chevron-shaped band of intense surface winds. The Norwegian Cyclone Model, as developed by the Bergen School of Meteorology, largely observed cyclones at the tail end of their lifecycle and used the term occlusion to identify the decaying stages.

Warm seclusions may have cloud-free, eye-like features at their center (reminiscent of tropical cyclones), significant pressure falls, hurricane-force winds, and moderate to strong convection. The most intense warm seclusions often attain pressures less than 950 millibars (28.05 inHg) with a definitive lower to mid-level warm core structure. A warm seclusion, the result of a baroclinic lifecycle, occurs at latitudes well poleward of the tropics.

As latent heat flux releases are important for their development and intensification, most warm seclusion events occur over the oceans; they may impact coastal nations with hurricane force winds and torrential rain. Climatologically, the Northern Hemisphere sees warm seclusions during the cold season months, while the Southern Hemisphere may see a strong cyclone event such as this during all times of the year.

In all tropical basins, except the Northern Indian Ocean, the extratropical transition of a tropical cyclone may result in re-intensification into a warm seclusion. For example, Hurricane Maria of 2005 re-intensified into a strong baroclinic system and achieved warm seclusion status at maturity (or lowest pressure).

Motion

Extratropical cyclones are generally driven, or "steered", by deep westerly winds in a general west to east motion across both the Northern and Southern hemispheres of the Earth. This general motion of atmospheric flow is known as "zonal". Where this general trend is the main steering influence of an extratropical cyclone, it is known as a "zonal flow regime".

When the general flow pattern buckles from a zonal pattern to the meridional pattern, a slower movement in a north or southward direction is more likely. Meridional flow patterns feature strong, amplified troughs and ridges, generally with more northerly and southerly flow.

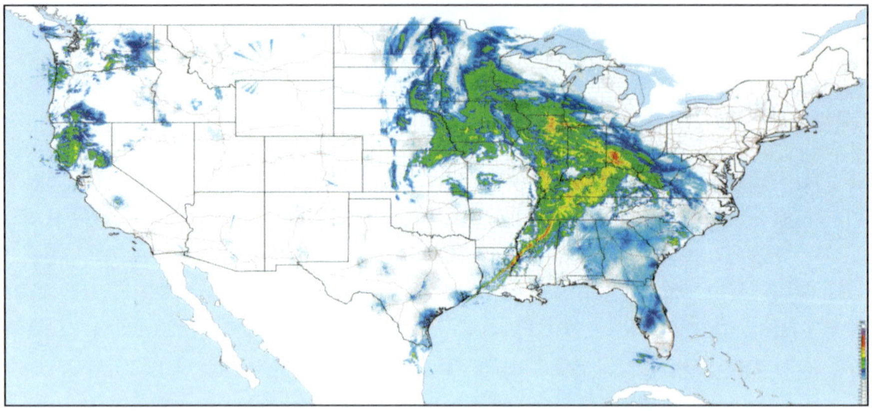

A February 24, 2007 radar image of a large extratropical cyclonic storm system at its peak over the central United States.

Changes in direction of this nature are most commonly observed as a result of a cyclone's interaction with other low pressure systems, troughs, ridges, or with anticyclones. A strong and stationary anticyclone can effectively block the path of an extratropical cyclone. Such blocking patterns are quite normal, and will generally result in a weakening of the cyclone, the weakening of the anticyclone, a diversion of the cyclone towards the anticyclone's periphery, or a combination of all three to some extent depending on the precise conditions. It is also common for an extratropical cyclone to strengthen as the blocking anticyclone or ridge weakens in these circumstances.

Where an extratropical cyclone encounters another extratropical cyclone (or almost any other kind of cyclonic vortex in the atmosphere), the two may combine to become a binary cyclone, where the vortices of the two cyclones rotate around each other (known as the "Fujiwhara effect"). This most often results in a merging of the two low pressure systems into a single extratropical cyclone, or can less commonly result in a mere change of direction of either one or both of the cyclones. The precise results of such interactions depend on factors such as the size of the two cyclones, their strength, their distance from each other, and the prevailing atmospheric conditions around them.

General Effects

Extratropical cyclones can bring mild weather with a little rain and surface winds of 15–30 km/h (9.3–18.6 mph), or they can be cold and dangerous with torrential rain and winds exceeding 119 km/h (74 mph), (sometimes referred to as windstorms in Europe). The band of precipitation that is associated with the warm front is often extensive. In mature extratropical cyclones, an area known as the comma head on the northwest periphery of the surface low can be a region of heavy precipitation, frequent thunderstorms, and thunder snows. Cyclones tend to move along a predictable path at a moderate rate of progress. During fall, winter, and spring, the atmosphere over continents can be cold enough through the depth of the troposphere to cause snowfall.

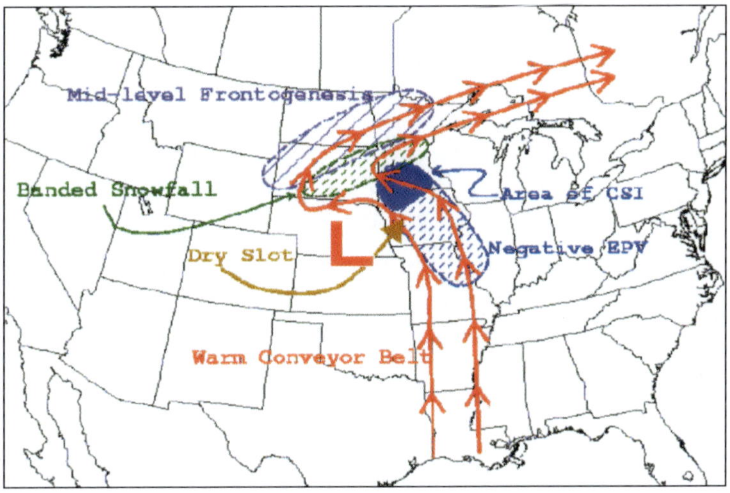

Preferred region of snowfall in an extratropical cyclone.

Severe Weather

Squall lines, or solid bands of strong thunderstorms, can form ahead of cold fronts and lee troughs due to the presence of significant atmospheric moisture and strong upper level divergence, leading to hail and high winds. When significant directional wind shear exists in the atmosphere ahead of a cold front in the presence of a strong upper level jet stream, tornado formation is possible. Although tornadoes can form anywhere on Earth, the greatest number occur in the Great Plains in the United States, because downslope winds off the north-south oriented Rocky Mountains, which can form a dry line, aid their development at any strength.

Explosive development of extratropical cyclones can be sudden. The storm known in Great Britain and Ireland as the "Great Storm of 1987" deepened to 953 millibars (28.14 inHg) with a highest recorded wind of 220 km/h (140 mph), resulting in the loss of 19 lives, 15 million trees, widespread damage to homes and an estimated economic cost of £1.2 billion (US$2.3 billion).

Although most tropical cyclones that become extratropical quickly dissipate or are absorbed by another weather system, they can still retain winds of hurricane or gale force. In 1954, Hurricane Hazel became extratropical over North Carolina as a strong Category 3 storm. The Columbus Day Storm of 1962, which evolved from the remains of Typhoon Freda, caused heavy damage in Oregon and Washington, with widespread damage equivalent to at least a Category 3. In 2005, Hurricane Wilma began to lose tropical characteristics while still sporting Category 3-force winds (and became fully extratropical as a Category 1 storm).

In summer, extratropical cyclones are generally weak, but some of the systems can cause significant floods overland because of torrential rainfall. The July 2016 North China cyclone never brought gale-force sustained winds, but it caused devastating floods in mainland China, resulting in at least 184 deaths and ¥33.19 billion (US$4.96 billion) of damage.

Climate and General Circulation

In the classic analysis by Edward Lorenz (the Lorenz energy cycle), extratropical cyclones (so-called atmospheric transients) acts as a mechanism in converting potential energy that is created by pole to equator temperature gradients to eddy kinetic energy. In the process, the pole-equator temperature gradient is reduced (i.e. energy is transported pole ward to warm up the higher latitudes).

The existence of such transients are also closely related to the formation of the Icelandic and Aleutian Low — the two most prominent general circulation features in the mid- to sub-polar northern latitudes. The two lows are formed by both the transport of kinetic energy and the latent heating (the energy released when water phase changed from vapor to liquid during precipitation) from the extratropical cyclones.

Mediterranean Tropical-like Cyclone

Mediterranean tropical-like cyclones, often referred to as medicanes (a portmanteau of Mediterranean hurricanes) but sometimes also as Mediterranean cyclones or as Mediterranean hurricanes, are meteorological phenomena observed over the Mediterranean Sea. On a few rare occasions, some storms have been observed reaching the strength of a Category 1 hurricane. The main societal hazard posed by Medicanes is not usually from destructive winds, but through life-threatening torrential rains and flash floods.

Satellite imagery of a well-documented Mediterranean tropical-like cyclone on 16 January 1995.

The occurrence of Medicanes has been described as not particularly rare. Tropical-like systems were first identified in the Mediterranean basin in the 1980s, when widespread satellite coverage showing tropical-looking low pressures which formed a cyclonic eye in the center were identified. Due to the dry nature of the Mediterranean region, the formation of tropical, subtropical cyclones and tropical-like cyclones are infrequent and also hard to detect, in particular with the reanalysis of past data. Depending on the search algorithms used, different long term surveys of satellite era and pre-satellite era data came up with 67 tropical-like cyclones of tropical storm intensity or higher between 1947–2014, and around 100 recorded tropical-like storms between 1947 and 2011. More consensus exists about the long term temporal and spatial distribution of tropical-like cyclones: they form predominantly over the western and central Mediterranean Sea while the area east of Crete is almost devoid of tropical-like cyclones. The development of tropical-like cyclones can occur year round, with activity historically peaking between the months of September and January, while the counts for the summer months of June and July are the lowest.

Climatology

Visible satellite imagery of a "medicane" above the Balearic Islands on 7 October 1996.

A majority of Mediterranean tropical cyclones (tropical cyclogenesis) form over two separate regions. The first, more conducive for development than the other, encompasses an area of the western Mediterranean bordered by the Balearic Islands, southern France, and the shorelines of the islands of Corsica and Sardinia. The second identified region of development, in the Ionian Sea between Sicily and Greece and stretching south to Libya, is less favorable for tropical cyclogenesis. An additional two regions, in the Aegean and Adriatic seas, produce fewer medicanes, while activity is minimal in the Levantine region. The geographical distribution of Mediterranean tropical-like cyclones is markedly different than that of other cyclones, with the formation of regular cyclones centering on the Pyrenees and Atlas mountain ranges, the Gulf of Genoa, and the island of Cyprus in the Ionian Sea. Although meteorological factors are most advantageous in the Adriatic and Aegean seas, the closed nature of the region's geography, bordered by land, allows little time for further evolution.

The geography of mountain ranges bordering the Mediterranean are conducive for severe weather and thunderstorms, with the sloped nature of mountainous regions permitting the development of convective activity. Although the geography of the Mediterranean region, as well as its dry air, typically prevents the formation of tropical cyclones, when certain meteorological circumstances arise, difficulties influenced by the region's geography are overcome. The occurrence of tropical cyclones in the Mediterranean Sea is generally extremely rare, with an average of 1.57 forming annually and merely 99 recorded occurrences of tropical-like storms discovered between 1948 and 2011 in a modern study, with no definitive trend in activity in that period. Few medicanes form during the summer season, though activity typically rises in autumn, peaks in January, and gradually decreases from February to May. In the western Mediterranean region of development, approximately 0.75 such systems form each year, compared to 0.32 in the Ionian Sea region. However, on very rare occasions, similar tropical-like storms may also develop in the Black Sea.

Studies have evaluated that global warming can result in higher observed intensities of tropical cyclones as a result of deviations in the surface energy flux and atmospheric composition, which both heavily influence the development of medicanes as well. In tropical and subtropical areas, sea surface temperatures (SSTs) rose 0.2 °C (0.36 °F) within a 50-year period, and in the North Atlantic and Northwestern Pacific tropical cyclone basins, the potential destructiveness and energy of storms nearly doubled within the same duration, evidencing a clear correlation between global warming and tropical cyclone intensities. Within a similarly recent 20-year period, SSTs in the

Mediterranean Sea increased by 0.6 to 1 °C (1.1 to 1.8 °F), though no observable increase in medicane activity has been noted, as of 2013. In 2006, a computer-driven atmospheric model evaluated the future frequency of Mediterranean cyclones between 2071 and 2100, projecting a decrease in autumn, winter, and spring cyclonic activity coinciding with a dramatic increase in formation near Cyprus, with both scenarios attributed to elevated temperatures as a result of global warming. In another study, researchers found that more tropical-like storms in the Mediterranean could reach Category 1 strength by the end of the 21st Century, with most of the stronger storms appearing in the autumn, though the models indicated that some storms could potentially reach Category 2 intensity. Other studies, however, have been inconclusive, forecasting both increases and decreases in duration, number, and intensity. Three independent studies, using different methodologies and data, evaluated that while medicane activity would likely decline with a rate depending on the climate scenario considered, a higher percentage of those that formed would be of greater strength.

Development and Characteristics

A Mediterranean tropical-like cyclone south of Italy.

The development of tropical or subtropical cyclones in the Mediterranean Sea can usually only occur under somewhat unusual circumstances. Low wind shear and atmospheric instability induced by incursions of cold air are often required. A majority of Medicanes are also accompanied by upper-level troughs, providing energy required for intensifying atmospheric convection—thunderstorms—and heavy precipitation. The baroclinic properties of the Mediterranean region, with high temperature gradients, also provide necessary instability for the formation of tropical cyclones. Another factor, rising cool air, provides necessary moisture as well. Warm sea surface temperatures (SSTs) are mostly unnecessary, however, as most Medicanes' energy are derived from warmer air temperatures. When these favorable circumstances coincide, the genesis of warm-core Mediterranean tropical cyclones, often from within existing cut-off cold-core lows, is possible in a conducive environment for formation.

Factors required for the formation of Medicanes are somewhat different than that normally expected of tropical cyclones; known to emerge over regions with sea surface temperatures (SSTs) below 26 °C (79 °F), Mediterranean tropical cyclones often require incursions of colder air to induce atmospheric instability. A majority of Medicanes develop above regions of the Mediterranean with SSTs of 15 to 26 °C (59 to 79 °F), with the upper bound only found in the southernmost reaches of the sea. Despite the low sea surface temperatures, the instability incited by cold atmospheric air within a baroclinic zone—regions with high differences in temperature and pressure—permits the formation of Medicanes, in contrast to tropical areas lacking high baroclinity, where raised

SSTs are needed. While significant deviations in air temperature have been noted around the time of Mediterranean tropical cyclones' formation, few anomalies in sea surface temperature coincide with their development, indicating that the formation of Medicanes is primarily controlled by higher air temperatures, not by anomalous SSTs. Similar to tropical cyclones, minimal wind shear—difference in wind speed and direction over a region—as well as abundant moisture and vorticity encourages the genesis of tropical cyclone-like systems in the Mediterranean Sea.

Satellite image of a tropical-like cyclone on December 15, 2005.

Due to the confined character of the Mediterranean and the limited capability of heat fluxes—in the case of Medicanes, air—sea heat transfer—tropical cyclones with a diameter larger than 300 km (190 mi) cannot exist within the Mediterranean. Despite being a relatively baroclinic area with high temperature gradients, the primary energy source utilized by Mediterranean tropical cyclones is derived from underlying heat sources generated by the presence of convection—thunderstorm activity—in a humid environment, similar to tropical cyclones elsewhere outside the Mediterranean Sea. In comparison to other tropical cyclone basins, the Mediterranean Sea generally presents a difficult environment for development; although the potential energy necessary for development is not abnormally large, its atmosphere is characterized by its lack of moisture, impeding potential formation. The full development of a Medicane often necessitates the formation of a large-scale baroclinic disturbance, transitioning late in its life cycle into a tropical cyclone-like system, nearly always under the influence of a deep, cut-off, cold-core low within the middle-to-upper troposphere, frequently resulting from abnormalities in a wide-spreading Rossby wave—massive meanders of upper-atmospheric winds.

A weak and disorganized Mediterranean tropical-like cyclone on 28 January 2009.

The development of Medicanes often results from the vertical shift of air in the troposphere as well, resulting in a decrease in its temperature coinciding with an increase in relative humidity, creating an environment more conducive for tropical cyclone formation. This, in turn, leads to in an increase in potential energy, producing heat-induced air-sea instability. Moist air prevents the occurrence of convective downdrafts—the vertically downward movement of air—which often hinder the inception of tropical cyclones, and in such a scenario, wind shear remains minimal; overall, cold-core cut-off lows serve well for the later formation of compact surface flux-influenced warm-core lows such as Medicanes. The regular genesis of cold-core upper-level lows and the infrequency of Mediterranean tropical cyclones, however, indicate that additional unusual circumstances are involved the emergence of the latter. Elevated sea surface temperatures, contrasting with cold atmospheric air, encourage atmospheric instability, especially within the troposphere.

In general, most Medicanes maintain a radius of 70 to 200 km (40 to 120 mi), last between 12 hours and 5 days, travel between 700 to 3,000 km (430 to 1,860 mi), develop an eye for less than 72 hours, and feature wind speeds of up to 144 km/h (89 mph); in addition, a majority are characterized on satellite imagery as asymmetric systems with a distinct round eye encircled by atmospheric convection. Weak rotation, similar to that in most tropical cyclones, is usually noted in a Medicane's early stages, increasing with intensity; Medicanes, however, often have less time to intensify, remaining weaker than most North Atlantic hurricanes and only persisting for the duration of a few days. The theoretical maximum potential intensity of Medicanes is equivalent to the lowest classification on the Saffir–Simpson hurricane wind scale, a Category 1 hurricane. While the entire lifetime of a cyclone may encompass several days, most will only retain tropical characteristics for less than 24 hours. Circumstances sometimes permit the formation of smaller-scale Medicanes, although the required conditions differ even from those needed by other Medicanes. The development of abnormally small tropical cyclones in the Mediterranean usually requires upper-level atmospheric cyclones inducing cyclogenesis in the lower atmosphere, leading to the formation of warm-core lows, encouraged by favorable moisture, heat, and other environmental circumstances.

Mediterranean cyclones have been compared to polar lows—cyclonic storms which typically develop in the far regions of the Northern and Southern Hemispheres—for their similarly small size and heat-related instability; however, while Medicanes nearly always feature warm-core lows, polar lows are primarily cold-core. The prolonged life of Medicanes and similarity to polar lows is caused primarily by origins as synoptic-scale surface lows and heat-related instability. Heavy precipitation and convection within a developing Mediterranean tropical cyclone are usually incited by the approach of an upper-level trough—an elongated area of low air pressures—bringing downstream cold air, encircling an existing low-pressure system. After this occurs, however, a considerable reduction in rainfall rates occurs despite further organization, coinciding with a decrease in previously high lightning activity as well. Although troughs will often accompany Medicanes along their track, separation eventually occurs, usually in the later part of a Mediterranean tropical cyclone's life cycle. At the same time, moist air, saturated and cooled while rising into the atmosphere, then encounters the Medicane, permitting further development and evolution into a tropical cyclone. Many of these characteristics are also evident in polar lows, except for the warm core characteristic.

Tornado

A tornado is a violently rotating column of air that is in contact with the base of a cumulonimbus cloud (or occasionally, a cumulus cloud) and the Earth's surface. Tornadoes come in many sizes, but they typically take the form of a visible condensation funnel whose narrow end touches the Earth and is often encircled by a cloud of debris.

Most tornadoes have wind speeds of 110 miles per hour (mph) (175 kilometers per hour (km/h)) or less, are approximately 250 feet (75 meters (m)) across, and travel a few miles (several kilometers) before dissipating. Some attain wind speeds of more than 300 mph (480 km/h), stretch more than a mile (1.6 km) across, and stay on the ground for dozens of miles (more than 100 km).

Tornadoes have been observed on every continent except Antarctica. Most, however, occur in the United States. They are also common in southern Canada, south-central and eastern Asia, east-central South America, Southern Africa, north-western and central Europe, Italy, western and south-eastern Australia, and New Zealand.

Tornado Occurrence and Distribution

Global Occurrence

Tornadoes have been reported on all continents except Antarctica. They are most common on continents in the mid-latitudes (between 20° and 60 °N and S), where they are frequently associated with thunderstorms that develop in regions where cold polar air meets warm tropical air.

Calculating which country has the most tornadoes per year depends on how this measurement is defined. The United Kingdom has the most tornadoes per land size, most of them weak. On average, about 33 tornadoes are reported annually there. In absolute numbers, the United States has the most tornadoes by far (more than 1,000 per year have been reported every year since 1990). It also has the most violent tornadoes (about 10 to 20 per year). Tornadoes of this intensity are very infrequent outside of the United States. Canada reports the second largest number of tornadoes (about 80 to 100 annually). Russia may have many tornadoes, but reports are not available to quantify their occurrence. About 20 tornadoes are reported in Australia each year, though the actual number is likely much higher. Many storms occur in uninhabited areas, and so any tornadoes that they produce are undocumented.

Records of tornado occurrences are fragmentary for many areas, making estimates of global tornado frequency difficult. Insurance records show that tornadoes have caused significant losses in Europe, India, Japan, South Africa, and Australia. Rare but deadly tornadoes have occurred in many other countries, including Bangladesh, China, and Argentina. There are few tornado reports from either the Arctic or the equatorial tropics.

In the United Kingdom almost all reported tornadoes are associated with vigorous convection occurring in advance of and along a cold frontal boundary. Large temperature differences are associated with early winter cold fronts that move rapidly across the country from the north and west, at times spawning widespread outbreaks of small tornadoes. For example, the passage of a very strong frontal boundary across the United Kingdom on November 23, 1981, produced 105 documented tornadoes. Similar phenomena occur in other European countries such as France and Belgium.

Most Southern Hemisphere tornadoes occur in Australia. Many reports come from New South Wales, where there were 173 reported tornadoes from 1901 to 1966. In addition, South Africa and Argentina both reported 191 tornadoes from 1930 to 1979. Because tornado formation is closely tied to the speed and directional shear of the wind with height, tornadoes in the Southern Hemisphere almost exclusively rotate clockwise, opposite to the rotation of their Northern Hemisphere counterparts.

Occurrence in the United States

Reported Events

Though tornadoes occur in every state, they are most frequent and attain the highest intensities in the central portion of the United States. Texas has the most reported tornadoes each year, about 125 on average for the years 1953–91; Florida, with almost 10 tornadoes per 10,000 square miles per year, has the most per area. However, most Florida tornadoes are very weak and affect extremely small areas.

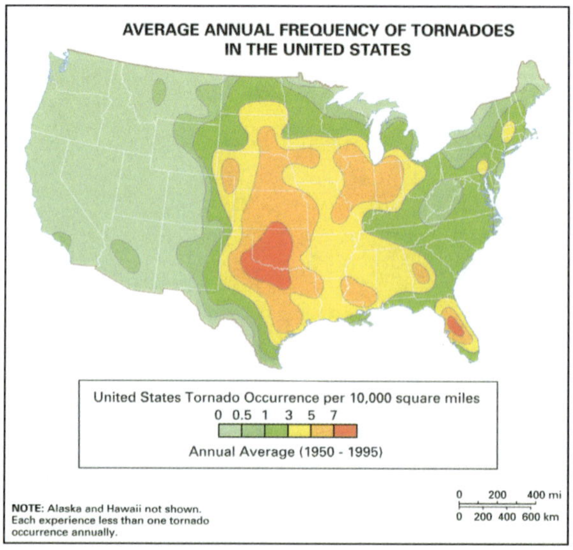

Map of the average annual frequency of tornadoes in the United States, showing the range of "Tornado Alley" from Texas through Nebraska.

From 1916 through 1998, about 45,000 tornadoes were documented in the United States. From 1916 to 1953, approximately 158 tornadoes were reported per year. After 1953, the beginning of the "modern period" of tornado documentation, the number of reports rose to more than 800 per year. (The modern period is considered to have begun in 1953 because this was the first full year in which the U.S. Weather Bureau issued tornado watches—that is, bulletins reporting that a tornado might be imminent.) The increased number of tornadoes reported was due to improvements in observing and recording (largely because of the establishment of a network of volunteer tornado "spotters"), which allowed a greater number of weak events to be recognized. During the interval from 1953 to 1998, there was an average of 169 "tornado days" (days on which one or more tornadoes were reported) per year. However, the early years of the modern period, with their relatively fewer reports, bias these averages. If only the 15 years 1984 through 1998 were considered, the average number of tornadoes per year would be 1,025, occurring on 173 tornado days per year. These higher numbers are credited to additional improvements in tornado reporting.

Geographic Distribution

When the number of tornado occurrences, their intensity, and the area they affect are considered, the centre of tornado activity is unquestionably seen to exist in the western portions of the southern Great Plains. The region of maximum tornado frequency, rightfully called Tornado Alley, extends from west Texas northeast through the western and central portions of Oklahoma and Kansas and across most of Nebraska.

Another area of frequent tornado occurrence is found across eastern Iowa, Illinois, Indiana, western Ohio, the southern portions of Wisconsin and Michigan, and the northern part of Kentucky. While this area experiences fewer tornadoes than does Tornado Alley, it has been struck by some of the strongest known tornadoes and has been the site of several large tornado outbreaks (that is, the occurrence of multiple tornadoes from the same weather system). The Gulf States (from east Texas to central Florida) have many weak tornadoes and have had outbreaks. The Gulf States also experience many tornadoes associated with hurricanes.

Seasonal Patterns

While most tornadoes develop in the spring and summer, tornadoes have occurred every day of the year. Several days have had many occurrences, reflecting large regional and national outbreaks. The distribution of reported tornadoes by month for the period 1916 through 1990 shows that about 74 percent of all tornadoes are reported from March through July. Peak months are April (14 percent), May (22 percent), and June (20 percent). December and January are the months of lowest activity.

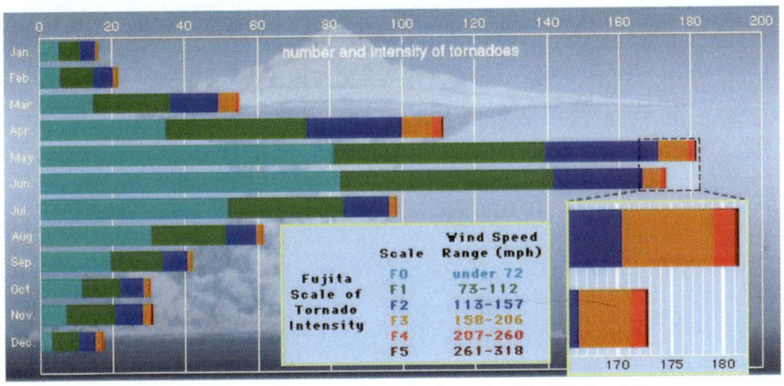

Graph of the number and intensity of tornadoes in the United States per month.

Figure indicates Tornado wind speed is ranked according to the Fujita Scale of tornado intensity. The occurrence of high-intensity tornadoes, though rare, is most common from March through June. Tornadoes are less common during the winter because air-mass boundaries are not as likely to be characterized by the strong temperature and moisture contrasts required to fuel powerful thunderstorms.

The main concentration of tornado activity migrates across the central portion of the United States in a seasonal cycle. Toward the end of winter (late February), the centre of tornado activity lies over the central Gulf States. At this time, southward-moving cold air reaches the southern limit of its expansion and encroaches on the Gulf Coast. As spring progresses, the days grow longer and more solar energy is intercepted. Land temperatures rise, and warm, moist air from the Gulf of Mexico progressively drives back the cold air. The centre of activity then moves eastward to the south-eastern Atlantic states, with tornado frequency peaking there in April.

As spring advances and gives way to summer, the centre of tornado activity gradually shifts westward and then northward. It moves across the southern Plains in May and June and then into the northern Plains and the Great Lakes states by early summer. Late summer through early fall is usually a relatively quiet time because the temperature and moisture contrasts across the boundary between the two air masses are weak. An extension of the Bermuda high (a centre of high atmospheric pressure that develops over the Atlantic Ocean) dominates the south-eastern third of the United States, and, while thunderstorms occur frequently in the warm, moist air, they seldom become severe. In late fall the days grow shorter, the temperature and moisture contrast intensifies again, and the centre of tornado activity retreats south toward the Gulf, completing the annual cycle.

Superimposed on this general pattern are large year-to-year variations. These arise because almost all tornado-producing storms are embedded within episodic northward surges of warm, moist air. The distribution of tornadoes in any one year thus reflects the weather patterns—especially the tracks followed by the synoptic-scale low-pressure centres—prevailing in that year.

Regional factors must also be taken into account. Along the Gulf Coast, tornadoes can be produced by thunderstorms that come ashore as a hurricane makes landfall. In a few cases, many tornadoes will be produced. For example, on September 20, 1967, thunderstorms in Hurricane Beulah produced 115 tornadoes in south Texas.

Diurnal Patterns

Although tornadoes can occur at all times of the day, they are most frequent from mid-afternoon to early evening. In the central United States, where most tornadoes occur, tornado frequency is highest between 5:00 and 6:00 PM. Weak and strong tornadoes occur most frequently in this same hour. Violent tornado occurrences peak an hour later, between 6:00 and 7:00 PM. Tornado occurrences peak in the late afternoon to early evening because there must be sufficient time for the Sun strongly to heat the ground and the surface layer of the atmosphere, thus inducing and sustaining severe thunderstorms. Few reports (roughly 1 percent) are from between 5:00 and 6:00 AM, the period just around sunrise when the atmosphere is often very stable.

Tornado Outbreaks

A tornado outbreak is the occurrence of several tornadoes over a region, usually due to thunderstorms embedded in the same synoptic-scale weather system. Outbreaks are classified according to the number of tornadoes reported: small (6 to 9 tornadoes), medium (10 to 19), and large (more than 20 tornadoes). Outbreaks are also classified according to the area affected: in local outbreaks, one or only a portion of one state is affected; in regional outbreaks, two or three states contain all or almost all the tornadoes; in national outbreaks, tornadoes are reported in many states.

On most tornado days, only a relatively small region has the essential conditions juxtaposed in just the right way to foster tornado development. This alignment usually lasts for only a small portion of an afternoon. As a consequence, only the two or three storms that form in this region are likely to produce tornadoes. Most of these vortices are weak, but one or more may be strong. Most occur in the space of one to two hours in the middle to late afternoon. Situations like this give rise to small local outbreaks several times per year.

Most dangerous are national outbreaks. Perhaps once every 10 to 15 years, the synoptic-scale weather pattern produces conditions favourable to the production of strong storms over a large portion of the central United States. The number of conditions required to align over this large an area usually limits the occurrence of such widespread outbreaks to March and April. Many of the tornadoes are likely to be strong or violent. The largest national tornado outbreak was the Super Outbreak of April 26–28, 2011, which spawned more than 300 tornadoes across the eastern United States. The second largest was the Super Outbreak of April 3–4, 1974, which was credited with producing 148 tornadoes in the central and southern United States (though 4 of these were later reclassified as downbursts by the meteorologist T. Theodore Fujita). The third largest was the April 11–12, 1965, Palm Sunday Outbreak.

Prediction and Detection of Tornadoes

The first step in predicting the likely occurrence of tornadoes involves identifying regions where conditions are favourable to the development of strong thunderstorms. Essential ingredients for the occurrence of such storms are cool, dry air at middle levels in the troposphere superimposed over a layer of moist, conditionally unstable air near the surface.

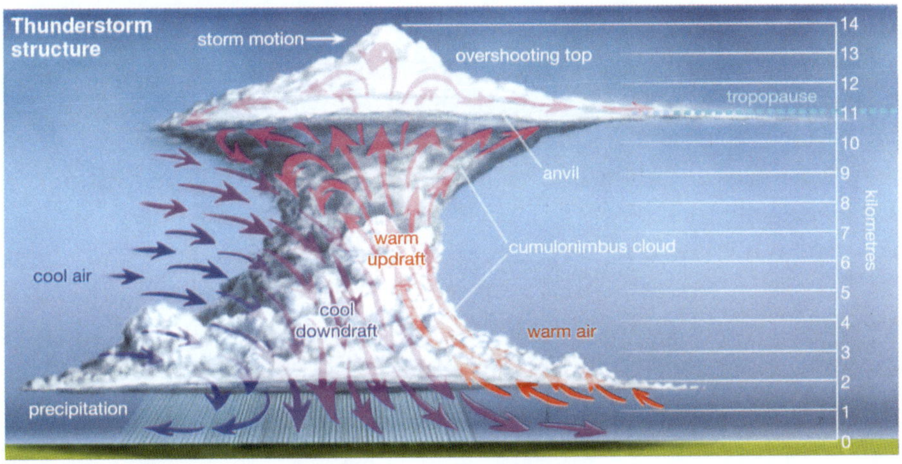

Structure of a thunderstorm.

Figure shows when the atmosphere becomes unstable enough to form large, powerful updrafts and downdrafts (as indicated by the red and blue arrows), a towering thundercloud are built up. At times the updrafts are strong enough to extend the top of the cloud into the tropopause, the boundary between the troposphere (or lowest layer of the atmosphere) and the stratosphere.

Conditions commonly leading to thunderstorm development occur along the warm side of the boundary line, or front that separates cold, dry air from warm, moist air. The degree of instability present in the atmosphere is approximated by the contrasts in temperature and moisture across the frontal boundary that divides the two air masses. For a storm to generate tornadoes, other factors must be present. The most important of these is a veering wind profile (that is, a progressive shifting of the wind, clockwise in the Northern Hemisphere, counter clockwise in the Southern Hemisphere, with increasing height) at low and middle levels, along with strong winds at high levels. Both of these wind actions are necessary to provide the required spin in the air that may eventually culminate in a tornado. A veering wind profile can be provided by the same strong temperature contrasts powering the thunderstorm, and high-altitude winds can be provided by the jet stream, the thin ribbon of high-speed air found in the upper half of the troposphere.

For the generation of a tornado, the diffuse spin must be concentrated into a small area as an evolving storm goes through several distinct stages of development. The first appearance of rotation in a storm is caused by the interaction of a strong, persistent updraft with the winds that blow through and around the storm. Rotation intensifies as the speed of the wind increases and as its direction veers from southeast to south and then around to west (in the Northern Hemisphere) with increasing height through the lower half of the troposphere.

Forecasters in the United States have learned to carefully monitor the wind profile in regions of instability and to estimate how temperatures and winds will evolve through the course of a day, while at the same time tracking the movement and intensity of the jet stream. With the aid of modern observing systems, such as vertically pointing radars (called wind profilers) and imaging systems on satellites that can measure the flow of water vapour through the Earth's atmosphere, forecasters can usually identify where conditions will be favourable for tornado formation one to seven hours in advance. This information is transmitted to the public as a tornado watch. A tornado warning is issued when a tornado has been spotted either visually or on weather radar.

Once strong thunderstorms begin to form, local offices of the National Weather Service monitor their development using imagery from satellite sensors and, most important, from radars. These allow forecasters to follow the evolution of the storms and to estimate their intensity. In the past, weather surveillance radars provided information only on the intensity of rainfall within the storms. Weather forecasters then had to infer the onset of rotation within a storm's updraft from circumstantial evidence, such as when the precipitation began to curve around the updraft to produce a "hook echo," a hook-shaped region of precipitation that flows out of the main storm and wraps around the updraft. Such inferences were highly subjective and prone to false alarms or very short-notice warnings. Today, modern weather surveillance radars not only provide information on the intensity of a storm's rainfall but also utilize the Doppler principle to sense winds within thunderstorms. Wind speeds are determined from radio waves reflected by raindrops and other particles carried along by the wind.

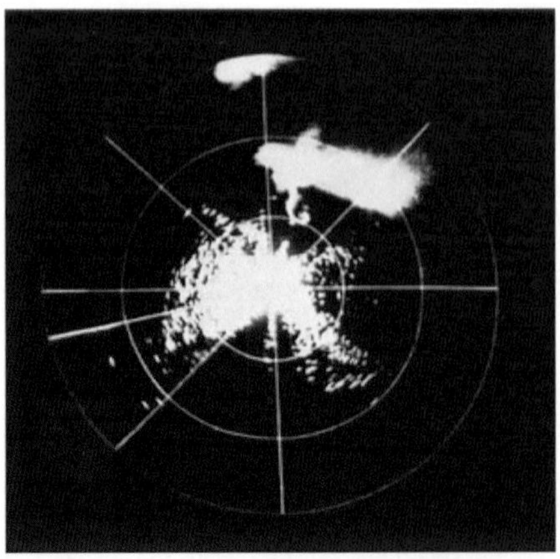

Hook echo of a tornado in Champaign, Ill., photographed on a radar scope on April 9, 1953. This was the first occasion on which the hook echo, an important clue in the tornado warning system, was recorded.

Doppler radars can measure rotation in the updraft and allow forecasters to watch the formation of a mesocyclone (that is, a region of rotating air within a thunderstorm). On Doppler radar, the presence of a well-organized mesocyclone is indicated by a small region of concentrated shear in the wind. On one side of the mesocyclone the rotating winds flow toward the radar; and on the other, they move away. In some cases, the formation of the tornado core can be detected. The tornado core is a roughly cylindrical region of lower atmospheric pressure that is bounded by the maximum tangential winds (the fastest winds circulating around the centre of the tornado). The radar indication of intense concentrated rotation is called the tornado vortex signature, although this area does not always evolve into a tornado core. These improvements have allowed forecasters to increase warning times while reducing false alarms.

Tornado-tracking activities under way with a field command vehicle from the National Severe Storms Laboratory (NSSL) in Goshen county, Wyo., as part of the Verification of the Origins of Rotation in Tornadoes Experiment 2 (VORTEX2), June 5, 2009.

Death and Damage

In the period 1916 through 1998, tornadoes claimed 12,282 lives in the United States, an average of 150 deaths per year. For the period 1953 through 1998, tornadoes claimed 4,032 lives, an average of 88 deaths per year. The decrease in fatalities is due to improvements in safety awareness and severe-weather warnings. There have been wide variations in the number of deaths from one year to the next, the minimum being 15 deaths in 1986 and the maximum being 805 in 1925 (due largely to the Great Tri-State Tornado of March 18, 1925). Since the advent of improved warning systems, three years contributed disproportionately to the total number of deaths: 1953, 1965, and 1974. Of the 519 deaths in 1953, 323 were due to strong tornadoes striking three urban areas (Waco, Texas; Flint, Michigan; and Worcester, Massachusetts). The high death counts in 1965 and 1974 (301 and 366, respectively) were the result of large national tornado outbreaks—the April 11–12, 1965, Palm Sunday Outbreak and the April 3–4, 1974, Super Outbreak.

Most deaths and injuries in tornadoes result from individuals being struck by flying debris. People are sometimes injured or killed by being rolled across the ground by the high winds. Also, a few people appear to have been killed by being carried aloft and then dropped from a great height.

Almost all of the damage caused by tornadoes can be attributed to wind-induced forces tearing structures apart. High-speed air flowing over a building's roof and around its corners produces forces that pull upward and outward, respectively. Once windows or doors on the windward side of a building break, air rushes in; this pushes outward on the remaining walls and upward on the roof from the inside, adding to the forces induced by the outside airflow. Often a building is torn apart so quickly and dramatically by these forces that it appears to explode.

It used to be thought that many buildings "exploded" owing to the development of an extreme difference in pressure between their interiors and the outside air. The rate at which pressure changes at a point on the surface as a tornado approaches may be as great as 100 hectopascals per second (100 hectopascals are equivalent to about 10 percent of atmospheric pressure at sea level). While this is a significant drop, studies have shown that most structures are sufficiently open that, even with the high rate of pressure change associated with a rapidly moving tornado, interior pressure can adjust quickly enough to prevent an explosion. Indeed, since the area of greatly reduced pressure beneath a tornado is small compared with the area of damaging winds, it is likely that a building already will have suffered damage from the winds by the time it experiences a rapid drop in outside pressure.

Tornado Safety

Tornadoes produce extremely hazardous conditions. The main dangers are caused by extremely high winds and flying debris. Little can be done to prevent heavy damage to structures that are directly hit by a tornado, though good building practices (such as securely fastening the roof of the house to the walls and securing the walls to the foundation) can reduce damage to structures on the periphery of a tornado's circulation. Consequently, most hazard research has been focused on saving lives and reducing injuries.

Surviving a tornado requires an understanding of the hazards and some preparation. The most important step is taking shelter. People are advised not to waste time opening windows to reduce the pressure in a house. It is a myth that sudden pressure changes associated with tornadoes cause structures to explode, and flying debris is likely to break the windows before the tornado hits anyway. The best protection in the home can be found under a sturdy table or workbench in the basement. In homes without basements (or when time is extremely limited), the recommendation is to shelter in a small interior room such as a closet or bathroom, preferably one with thick walls and no windows. A mattress can be used as additional cover, and a heavy blanket can protect against dust. It is recommended that people also shield their head and neck with their arms. People are strongly advised to avoid windows in all cases, as flying glass can cause terrible wounds.

Many of the fatalities produced by tornadoes occur in mobile homes. Such homes are very lightly constructed and begin to come apart at relatively low wind speeds. In addition, they can easily be blown over and disintegrate, producing ample amounts of sharp-edged flying debris. Consequently, mobile homes—even ones with anchors or tie-downs—offer no shelter from even very weak tornadoes. At the first indication of possible tornadic activity, residents of mobile homes are advised to seek shelter in sturdy buildings. If no such building is available, it is better to shelter in a ditch or culvert than to remain in the home.

If caught in the open in tornadic conditions, it is recommended to stay low to the ground (preferably in a ditch or culvert) and to hold onto a sturdy object such as a tree stump. The main danger is from being tumbled along the ground by high winds, which can in effect beat a person to death.

It is advised that cars be abandoned as a tornado approaches and that shelter be sought in a ditch. Many tornado-related deaths have occurred in traffic jams, such as those associated with the April 10, 1979, tornado in Wichita Falls, Texas. Cars tend to be tumbled over and over by tornadoes and, in extreme cases, carried aloft and dropped from significant heights. Victims are often thrown from the tumbling car or blown out through the windows.

Tornado Intensity

Tornado intensity is not estimated directly from measured wind speeds, because tornadoes rarely pass near meteorological instruments. Rather, it is commonly estimated by analyzing damage to structures and then correlating that damage with the wind speeds required to produce such destruction. This method is essential to assigning tornadoes specific values on the Enhanced Fujita Scale, or EF-Scale, of tornado intensity. The notion of developing such a scale for use in comparing events and in research was proposed in 1971 by the Japanese American meteorologist T. Theodore Fujita.

Fujita's scale was widely used in the United States and adapted for use in other parts of the world; however, almost from the beginning, the limitations of his approach were recognized. The primary criticisms were a lack of sufficient damage indicators for the many building types found in modern society, no recognition of regional variations in types and quality of construction of otherwise similar structures, and a lack of a definitive correlation between observed damage and wind speed. As a result, there were inconsistencies in the rating of tornado intensity in the historical records (which is shown as noise in statistical analyses). Also, tornado wind speeds, especially in very high wind events, were overestimated.

In 2004, after 33 years of experience with the original Fujita Scale, leading atmospheric researchers and tornado forecasters developed a plan to improve the estimation process and eliminate some of the limitations. The result, the Enhanced Fujita Scale, was adopted for use in 2007. It retains many of the features of the original scale but provides more precision at the higher intensity values. The scientists and forecasters also worked out ways to adjust the older records so that EF values would be available for the comparison.

To classify a tornado using the EF-Scale, the damage occurring along the tornado's track is mapped. A tornado's intensity varies along its path, with the most extreme damage usually being restricted to a small area, and the overall EF-Scale value assigned represents the tornado's highest attained intensity. Once the degree of damage to structures and vegetation has been assessed and matched with the appropriate EF-Scale value, the maximum wind velocities within the range associated with the EF-Scale value are assigned. Even in its improved form, such a system is inevitably limited. For example, a powerful tornado that does not pass near buildings or trees, causing little or no damage, may be given an EF-Scale value less than its true intensity.

The Enhanced Fujita Scale recognizes tornadoes of six different intensities ranging in number from EF0 to EF5. For many purposes, these can be grouped into three broader categories—weak, strong, and violent.

Weak EF0 and EF1 Tornadoes

Though most tornadoes (60 to 70 percent) are in this category, they account for less than 5 percent of all deaths. A weak tornado usually has a single funnel cloud (that is, a column of water droplets) resembling an elongated, upward-opening cone with a smooth surface. The cone often does not touch the ground. In weak tornadoes, vertical wind speeds are thought to be greatest along the central axis of circulation. Many weak tornadoes appear not to extend upward far beyond the base of the parent storm.

Machine shed pushed from its foundation, the type of "moderate damage" associated with weak tornadoes (ranking F1 on the Fujita Scale of tornado intensity).

Strong EF2 and EF3 Tornadoes

About 35 percent of all tornadoes are in the strong category, and they account for about 30 percent of all deaths. Typically, a strong tornado has a broad, columnar funnel cloud. The funnel surface usually has a rough, rapidly changing texture, reflecting small-scale turbulence. Available evidence suggests that in a strong tornado, most of the rising air surges upward in a cylindrical annulus around the central axis. Vertical speeds are lower along the axis itself. Sometimes "suction vortices" can be seen within the tornado core at its point of contact with the ground. This little-understood feature appears to contain the highest wind speeds in the tornado. Strong tornadoes extend well up into the generating thunderstorm because they generally form in or around a strongly rotating updraft that may persist through the storm's full height.

Multiunit building with its roof and many walls destroyed, the type of "severe damage" associated with strong tornadoes (ranking F3 on the Fujita Scale of tornado intensity).

Violent EF4 and EF5 Tornadoes

Only a very few tornadoes (2 percent or so) reach intensities high enough to be categorized as violent; however, they account for about 65 percent of all deaths. In many cases, a violent tornado has a broad core with a diameter of 0.5 km (0.3 mile) or more. At the centre of the core, there is a relatively calm and clear eye. In the eye, nonswirling air flows down from upper levels of the thunderstorm due to low pressure in the base of the core. Upon reaching the ground, this descending inner flow turns outward and mixes with air rushing in from the inflow boundary layer (that is, the layer of air near ground level). The combined flow then spirals upward around the eye in an annulus.

Completely disintegrated residential subdivision, the type of "incredible damage" associated with the most violent tornadoes (ranking F5 on the Fujita Scale of tornado intensity).

In some violent tornadoes, secondary vortices may form in the annulus, giving rise to what is termed a multiple-vortex tornado. In these secondary vortices, air spins rapidly around the axes while the vortices themselves rotate around the periphery of the central eye. Small secondary vortices are also called suction vortices when they are most evident in the corner region, the area where the wind entering the base of the tornado abruptly "turns the corner" from primarily horizontal to vertical flow. A tornado with one or more suction vortices is distinguished from a multiple-vortex tornado in that a suction vortex is at most only several hundreds of metres high, while multiple vortices extend all the way up into the cloud base of the parent thunderstorm. The fastest known surface winds occur around the tips of secondary vortices.

Physical Characteristics of Tornadoes

Airflow Regions

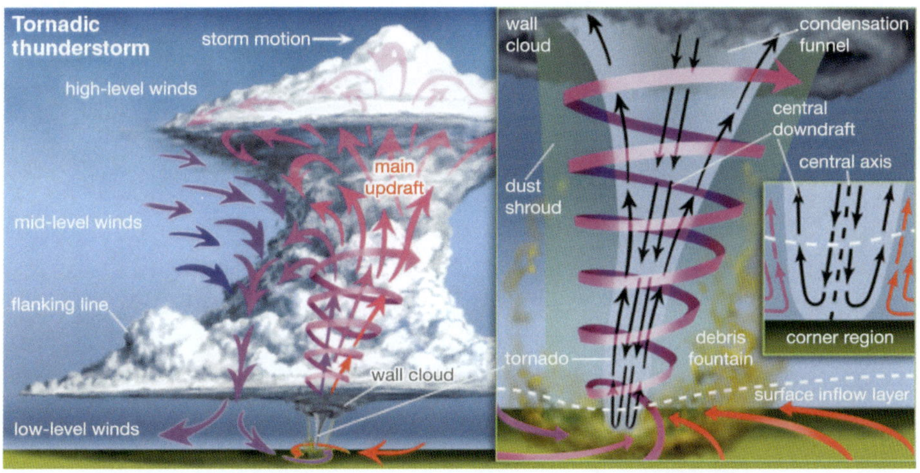

In figure, Tornadic thunderstorm (Left): The rotating updraft that produces the tornado extends high into the main body of the cloud. Anatomy of a tornado (Right): Air feeds into the base of a tornado and meets the tornado's central downflow. These flows mix and spiral upward around the central axis. The tornado's diameter can be much greater than that of the visible condensation funnel. At times the tornado may be hidden by a shroud of debris lifted from the ground.

Fully developed tornadoes contain distinct regions of airflow. As is shown in the figure, the central axis of circulation is within the core region, a roughly cylindrical area of lower atmospheric pressure that is bounded by the maximum tangential winds (the fastest winds circulating around the centre of the tornado). If a visible funnel cloud forms, it will occur within the core region. The funnel cloud consists of a column of water droplets, commonly called the condensation funnel. In very dry conditions there may be no condensation funnel associated with a tornado.

Responding to the reduced pressure in the central core, air near the ground located in what is referred to as the inflow boundary layer converges from all directions into a tornado's "corner region." This region gets its name because the wind abruptly "turns the corner" from primarily horizontal to vertical flow as it enters the core region and begins its upward spiral. The corner region is very violent. It is often marked by a dust whirl or a debris fountain, where the erupting inflow carries aloft material ripped from the surface. The inflow boundary layer that feeds the corner region is usually a few tens of metres deep and has turbulent airflow. Above the boundary layer, the core is surrounded by a weakly swirling outer flow—the inflow to the storm's updraft—where radial motions (movements toward or away from the tornado's axis) is relatively small. Somewhere aloft (exactly where is not known), the core and the swirling outer flow merge with the updraft of the generating thunderstorm.

Winds in a tornado are almost always cyclonic; that is, they turn counter clockwise in the Northern Hemisphere and clockwise in the Southern Hemisphere. This dominance of rotation direction is indirectly due to the Earth's rotation, which plays a role in controlling the structure of all large-scale weather systems. Most tornadoes are produced by thunderstorms, and a tornado's parent thunderstorm is in turn embedded within a larger weather system that determines the vertical shear in the winds (that is, their change in speed and direction with height across the troposphere). These systems rotate cyclonically, and a tornado's rotation comes from a concentration of the spin present in the sheared winds. However, not all tornadoes are cyclonic. About 5 percent of all observed tornadoes rotate anti-cyclonically—that is, they turn clockwise in the Northern Hemisphere and counter clockwise in the Southern Hemisphere.

Wind Speeds and Air Pressures

Measurement of wind speeds can be obtained by photogrammetry (measurements from photographs) and through remote sensing techniques using the Doppler Effect. These two techniques are complementary. They provide information about tornado wind speeds by tracking objects in and around the core (the assumption being that the objects are moving with the speed of the air). Photogrammetry allows speeds across the image plane to be determined by analysis of motions of dust packets, pieces of vegetation, and building debris as recorded on film or videotape, but it cannot be used to determine wind speed toward or away from the camera. On the other hand, through processing of Doppler-shifted electromagnetic "echoes" received from raindrops and debris illuminated with pulses of radio waves (radar) or light (lidar), wind speed toward or away from the instrument can be determined.

Under some conditions, extreme wind speeds can occur in the corner region of a tornado. The few measurements of violent tornado winds that have been made using Doppler radar and photogrammetry suggest that the maximum possible tangential wind speeds generated by tornadoes are in the range of 125 to 160 metres per second, or 450 to 575 km per hour (about 410 to 525 feet per second, or 280 to 360 miles per hour). Most researchers believe the actual extreme value is near the lower end of this range. Consistent with this thinking was the measurement made using a mobile Doppler radar of the fastest wind speed ever measured, 318 miles per hour (about 512 km per hour), in a tornado that hit the suburbs of Oklahoma City, Oklahoma, on May 3, 1999.

Maximum tangential speeds occur in a ring-shaped region that surrounds the tip of the vortex core that is centred 30 to 50 metres (100 to 160 feet) above the ground. (Hence, they tend to be a bit higher than damage-causing winds at the surface.) The vertical speeds of air rising as a central jet through the hole in the ring may be as high as 80 metres per second, or 300 km per hour (about 250 feet per second, or 170 miles per hour). Radial speeds of air flowing from the inflow region to the corner region (which feeds the central jet) are estimated to reach 50 metres per second, or 180 km per hour (about 160 feet per second, or 110 miles per hour). Because the organization of the airflow varies considerably with tornado intensity, extremes in vertical and radial speeds may not occur at the same time as extremes in tangential speeds.

These extreme speeds are the strongest winds known to occur near the Earth's surface. In reality, they occur over a very small portion of the tornado core close to the ground. Their actual occurrence is rare, and, when they do occur, they usually last only a very short time.In almost all tornadoes (about 98 percent), the maximum attained wind speed is much less than these maximum possible speeds.

While there have not been any direct measurements of atmospheric pressure in tornadoes, a few measurements have been taken when tornadoes passed near weather stations with barographs (instruments that record atmospheric pressure over time). Data from such incidents, along with measurements made in laboratory vortices, provide for the construction of mathematical models describing the distribution of surface pressure beneath tornadoes. These models, combined with information on tornado winds, are used to extrapolate what was the most likely air pressure at the centre of any given tornado.

These extrapolations indicate that a region of low surface pressure is centred beneath the tornado core. The area of this region is relatively small compared with that of the annulus of high-speed winds that surrounds it. Even for violent tornadoes, the reduction in surface pressure in this area (relative to surface pressure in the surrounding atmosphere) is probably no more than 100 hectopascals (that is, about 10 percent of standard atmospheric pressure at sea level). In most tornadoes, the reduction in central surface pressure is not that great.

The lowest atmospheric pressure in a tornado is thought to be at the centre of the core a few tens to a few hundred metres above the surface, though the magnitude of the pressure reduction is unknown. In violent tornadoes this pressure difference appears to be sufficient to induce a central downflow.

Funnel Clouds

A tornado is often made visible by a distinctive funnel-shaped cloud. Commonly called the condensation funnel, the funnel cloud is a tapered column of water droplets that extends downward from the base of the parent cloud. It is commonly mixed with and perhaps enveloped by dust and debris lifted from the surface. The funnel cloud may be present but not visible due to heavy rain. Over a tornado's lifetime, the size and shape of the funnel cloud may change markedly, reflecting changes in the intensity of the winds, the moisture content of the inflowing air, properties of the ground, and other factors. Very frequently the condensation funnel extends from the parent cloud only partway to the ground, and in very dry conditions there may be no condensation funnel. Generally, the more moist the air and the more intense the tornado, the larger the funnel cloud.

The funnel cloud usually outlines only the innermost core. Typically, its diameter is at most one-tenth that of the overall tornado circulation. Indeed, a tornado can occur without a funnel cloud being present at all. The funnel cloud's length can range from tens of metres to several kilometres; its diameter can span a few metres to hundreds of metres. Funnel clouds of weak tornadoes are usually cone-shaped, while strong and violent tornadoes form short, broad, cylindrical pillars. Long, rope-like tubes that trail off horizontally are common in the waning phase of many tornadoes.

Duration

The lifetime of a tornado is directly related to its intensity, with more intense tornadoes tending to last longer. On average, a tornado is on the ground for about 15 minutes, but this value is misleading because the average is heavily weighted by the rare but long-lived violent tornadoes. Most tornadoes are weak, lasting only about two to three minutes on average. A typical lifetime for strong tornadoes is about 8 minutes, while for violent events it is about 25 minutes. In exceptional cases, violent events can last more than three hours.

Speed and Direction of Movement

The movement of a tornado is determined by the motion of the generating thunderstorm. The average tornado moves at a speed of about 12 to 13 metres per second, or 43 to 47 km per hour (about 39 to 43 feet per second, or 27 to 29 miles per hour), but some have remained nearly stationary while others have traveled faster than 25 metres per second, or 90 km per hour (80 feet per second, or 55 miles per hour). As an extreme example, speeds of up to 33 metres per second, or 120 km per hour (110 feet per second, or 75 miles per hour) were measured in a tornado that struck Guin, Alabama, on April 3, 1974.

Most tornado-producing thunderstorms occur in a warm air mass that is under the influence of an active synoptic-scale low-pressure system (such a system covers about one-half of the continent). The middle-level winds (3 to 10 km [2 to 6 miles] in altitude) that in large part determine the direction of storm motion tend to be from the west or southwest in the Northern Hemisphere. Hence, most tornadoes (around 80 percent) come from the west or southwest and move to the east or northeast. Tornadoes move from northwest to southeast about 5 percent of the time. Many hurricane-related tornadoes have traveled east to west, as have a few Great Plains and Midwest tornadoes. In the Southern Hemisphere, storms (and consequently tornadoes) tend to move from the west or northwest to the east or southeast.

Tornado Cyclones, Tornado Families and Long-track Tornadoes

About 90 percent of tornadoes are associated with thunderstorms, usually supercells; this association accounts for many weak and almost all strong and violent tornadoes. The other 10 percent of tornado occurrences are associated with rapidly growing cumulus clouds; these vortices are almost always weak and short-lived.

As a very rough estimate, about 100,000 thunderstorms occur in the United States each year. About 10 percent of these (or about 10,000 per year) will become severe thunderstorms, and only about 5 percent to 10 percent of these severe storms (or about 500 to 1,000 per year) will produce tornadoes.

The typical tornado-producing thunderstorm lasts for two to three hours and usually produces one or two relatively short-lived tornadoes. The period of storm maturity during which a tornado is most likely to form may last only a few tens of minutes. However, on rare occasions a storm may produce a tornado cyclone (a core of concentrated rotation within the storm from which tornadoes are spawned) that is stable and long-lived. The strength of the tornado cyclone usually pulsates, creating a sequence of tornadoes. This gives rise to what is known as a tornado family. Tornado families typically have two or three members, though they can be much larger. During the Super Outbreak of April 3–4, 1974, in the United States, a single storm traveling along the Ohio River produced a family with eight members spread over several hundred kilometres.

On very rare occasions, the strength of a tornado cyclone will remain nearly constant for several hours, forming a single, long-lasting tornado with a continuous damage path many times the average length. This is referred to as a long-track tornado. Long-track tornadoes can be difficult to distinguish from tornado families. For instance, the Great Tri-State Tornado of March 18, 1925, is credited with a path length of 352 km (219 miles), though it cannot be proved that this event, which affected Missouri, Illinois, and Indiana, was an individual tornado or a series in the same family. On the other hand, the Monticello, Indiana, tornado on April 3, 1974 (part of the Super Outbreak mentioned above), produced a continuous track of damage for over 160 km (99 miles). It was also the fifth and final member of a tornado family.

Tornado Formation

Mesocyclone

Tornadoes may occur wherever conditions favour the development of strong thunderstorms. Essential conditions for such storms are the presence of cool, dry air at middle levels in the troposphere, overlying a layer of moist, conditionally unstable air near the surface of the Earth. Conditional instability occurs when a saturated air parcel (air at 100 percent relative humidity) continues to rise once set in motion, but an unsaturated air parcel resists being displaced vertically. The unsaturated air, if moved upward, will be cooler than the surrounding air and it will sink. On the other hand, when conditionally unstable air rises it becomes warmer owing to the condensation of water vapour. As the water condenses, heat is released, further warming the air and fueling its rise. This convective action (that is, the circulation of air as a result of heat transfer) produces the huge clouds commonly associated with thunderstorms and tornadoes. Convection can be initiated when the Sun heats a localized area of the ground, destabilizing the near-surface air.

Thunderstorms can also form along the boundary, or front, between air masses of different temperatures. In this case, the denser cool air displaces the warmer and forces it to rise. The greater the contrast in temperature and moisture across the frontal boundary, the greater the instability of the atmosphere and the greater the likelihood of a strong thunderstorm.

Most tornadoes are formed when a strong updraft such as those described above acts to concentrate atmospheric rotation, or spin, into a swirling column of air. Spin is a natural occurrence in air because horizontal winds almost always experience both an increase in speed and a veering in direction with increasing height above the surface. The increase of wind speed with height (called vertical speed shear) produces "crosswise spin," that is, rotation about a horizontal axis crosswise to the direction of wind flow. When air containing crosswise spin flows into an updraft, the spin is drawn upward, producing rotation about a vertical axis. The veering of wind direction with height (vertical direction shear) is another source of horizontal spin, this time oriented in the same direction as the wind flow and known as "streamwise spin." When air containing streamwise spin is drawn into an updraft, it too is tilted upward and rotates about a vertical axis. Although crosswise spin and streamwise spin are oriented at right angles to each other, both rotations exist in the horizontal plane, and both types have been revealed by Doppler radar observations to contribute to the evolution of a rotating updraft. Radar observations also have shown that updraft rotation makes its appearance in a thunderstorm at altitudes of 4 to 8 km (2.5 to 5 miles). At first, the tilting of crosswise spin into the vertical appears to be the principal mechanism of rotation; subsequently, as updraft rotation intensifies, the tilting of streamwise spin becomes more important. The resulting swirling column of rising air, perhaps 10 to 20 km (6 to 12 miles) in diameter and only weakly rotating, is called a mesocyclone.

Dynamic Pipe

As spin-up of the mesocyclone continues, its rotating action begins to reorganize airflow in the updraft. The local pressure field and the strongly curved wind field move toward a dynamic equilibrium called cyclostrophic balance. In this state, the pressure-gradient force, which acts to move air inward in response to the lower pressure in the centre of the rotating column, is equaled by the outward-directed centrifugal force. When cyclostrophic balance is achieved, air readily flows in a circular path around the mesocyclone's axis, while flow toward or away from its centre is strongly suppressed. This state, in which airflow is constrained by its own rotation, is known as the dynamic pipe effect.

The middle level of the storm is usually the first area where cyclostrophic balance is achieved, and the mesocyclone that begins to act as a dynamic pipe. Almost all the air flowing along the mesocyclone's axis is drawn in through the bottom of the pipe. This inflow further intensifies rotation at the pipe's lower end, causing it to extend rapidly downward as the more quickly rotating region comes into cyclostrophic balance.

Strong convergence of inflowing air at the lower end of the pipe causes air parcels to be accelerated upward and vertically "stretched." Vertical stretching normally causes the mesocyclone to contract to a diameter of about 2 to 6 km (1 to 4 miles). As this happens, the mesocyclone rotates more quickly, which in turn strengthens the convergence of inflowing winds at its base. In this manner the mesocyclone grows in strength in a positive-feedback, or self-amplifying, process.

Development of the dynamic pipe effect can produce a mesocyclone that extends the full depth of the thunderstorm, from about 1 km (0.6 mile) above the ground to near the storm's top at about 15 km (9 miles). Frequently, the maturation of the mesocyclone is heralded at the bottom of the cloud by a lowering of a portion of the thunderstorm's base in the area of the updraft. This approximately cylindrical extension is known as a wall cloud. Surface winds with speeds as high as 33 metres per second, or 120 km per hour (110 feet per second, or 75 miles per hour) can be present beneath this swirling cloud, often producing damage even when no tornado forms.

Tornado Core and the Condensation Funnel

The extension of a concentrated swirling core to the surface—in other words, the actual formation of a tornado—can occur once the mesocyclone is established. Most mesocyclones do not generate tornadoes. In the ones that do, a small region of increased convergence and stretching that is typically no more than one kilometre in diameter develops in the mesocyclone for reasons that have so far eluded storm researchers. This usually occurs at the interface between the thunderstorm's updraft and downdraft. Enhanced spin begins several kilometres above the ground, and then quickly builds downward. Around such a small volume, rotation is strong enough for a smaller dynamic pipe to form and extend to within several tens of metres of the surface. This dynamic pipe is called the tornado core. Once it forms, the parent mesocyclone is reclassified as a tornado cyclone.

As the core approaches the ground, surface friction slows the rotational motion and prevents the establishment of cyclostrophic balance. Surface friction also limits the rate of airflow into the base of the core. This restriction prevents inflow from filling the tornado's low-pressure core from below. At the same time, the parent storm's strong updraft prevents sufficient air from filling the core from above.

With air pressure in the vortex core thus reduced in comparison to the pressure outside the core at the same elevation, a condensation funnel forms. This occurs because, at lower atmospheric pressure, air flowing upward in the core cools more quickly with increasing height than air rising at higher pressure just outside the core. Assuming that inflowing air has the same amount of moisture throughout, air rising in the core reaches its dew point at a lower height than air rising just outside the core; any further rise leads to condensation and a visible cloud. Because pressure is lowest at the axis of the vortex, air rising along this centre line reaches its dew point nearer the ground than air spiraling up just a short distance outward. These processes gives rise to the characteristic conical or funnel shape of the condensation cloud.

Location in the Parent Storm

Many weak tornadoes form between the surface and the lowest portion of the parent cloud. These tornadoes exist for only a short time (a few minutes). Such tornadoes most commonly form beneath the flanking line of cumulus congestus clouds that frequently develop above a strong thunderstorm's gust front (the leading edge of the storm's downdraft). Often called gustnadoes, these vortices are true tornadoes when they are attached to the updraft of a rapidly growing congestus cloud. Gustnadoes draw their spin from the wind shear across the gust front. Their transient nature, relatively small diameters, and lack of a rotating region within the generating cloud cause them to be difficult to observe with radar. As a result, small tornadoes are not well documented, and in many respects they are less understood than stronger events.

In contrast to gustnadoes, almost all strong and violent tornadoes (and some weak ones as well) are closely connected with a rotating updraft that extends through much of the height of the parent storm. Such tornadoes tend to form near the interface between a storm's updraft and downdraft. To an observer on the ground, they are generally perceived as being beneath the right-rear quadrant of the main body of the storm in the Northern Hemisphere as viewed along the storm path. Because of the connection of these tornadoes to large-diameter circulations within the thunderstorm, many of the events leading to their formation have been fairly well documented both visually and by radar.

Acid Rain

Acid rain is a result of air pollution. When any type of fuel is burnt, lots of different chemicals are produced. The smoke that comes from a fire or the fumes that come out of a car exhaust don't just contain the sooty grey particles that you can see - they also contains lots of invisible gases that can be even more harmful to our environment.

Power stations, factories and cars all burn fuels and therefore they all produce polluting gases. Some of these gases (especially nitrogen oxides and sulphur dioxide) react with the tiny droplets of water in clouds to form sulphuric and nitric acids. The rain from these clouds then falls as very weak acid - which is why it is known as "acid rain".

Acidity of Acid Rain

Acidity is measured using a scale called the pH scale. This scale goes from 0 to 14. 0 is the most acidic and 14 is the most alkaline (opposite of acidic). Something with a pH value of 7, we call neutral, this means that it is neither acidic nor alkaline.

Very strong acids will burn if they touch your skin and can even destroy metals. Acid rain is much, much weaker than this, never acidic enough to burn your skin.

Rain is always slightly acidic because it mixes with naturally occurring oxides in the air. Unpolluted rain would have a pH value of between 5 and 6. When the air becomes more polluted with nitrogen oxides and sulphur dioxide the acidity can increase to a pH value of 4. Some rain has even been recorded as being pH2.

Vinegar has a pH value of 2.2 and lemon juice has a value of pH2.3. Even the strongest recorded acid rain is only about as acidic as lemon juice or vinegar and we know that these don't harm us - so why do we worry about acid rain.

Effects of Acid Rain

Acid rain can be carried great distances in the atmosphere, not just between countries but also from continent to continent. The acid can also take the form of snow, mists and dry dusts. The rain sometimes falls many miles from the source of pollution but wherever it falls it can have a serious effect on soil, trees, buildings and water.

Forests all over the world are dying, fish are dying. In Scandinavia there are dead lakes, which are crystal clear and contain no living creatures or plant life. Many of Britain's freshwater fish are threatened; there have been reports of deformed fish being hatched. This leads to fish-eating birds and animals being affected also. Is acid rain responsible for all this? Scientists have been doing a lot of research into how acid rain affects the environment.

Forests

It is thought that acid rain can cause trees to grow more slowly or even to die but scientists have found that it is not the only cause. The same amount of acid rain seems to have more effect in some areas than it does in others.

As acid rain falls on a forest it trickles through the leaves of the trees and runs down into the soil below. Some of it finds its way into streams and then on into rivers and lakes. Some types of soil can help to neutralise the acid - they have what is called a "buffering capacity".

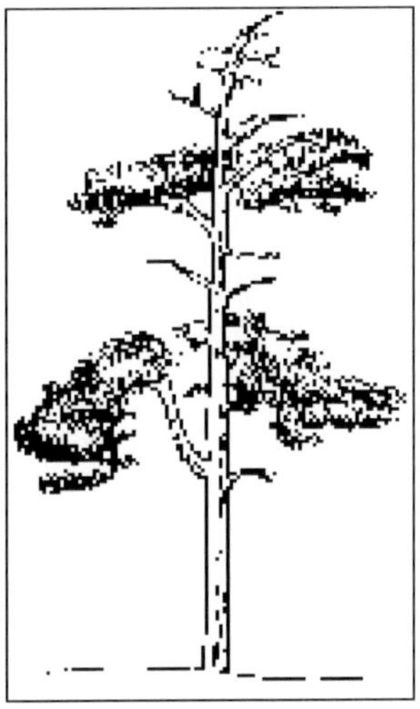

Other soils are already slightly acidic and these are particularly susceptible to the effects of acid rain. Acid rain can affect trees in several different ways, it may:

- Dissolve and wash away the nutrients and minerals in the soil which help the trees to grow.
- Cause the release of harmful substances such as aluminium into the soil.
- Wear away the waxy protective coating of leaves, damaging them and preventing them from being able to photosynthesise properly.

A combination of these effects weakens the trees which mean that they can be more easily attacked by diseases and insects or injured by bad weather. It is not just trees that are affected by acid rain, other plants may also suffer.

Lakes and Rivers

The acidity of the water does not just affect species directly; it also causes toxic substances like aluminium to be released into the water from the soil, harming fish and other aquatic animals.

It is in aquatic habitats that the effects of acid rain are most obvious. Acid rain runs off the land and ends up in streams, lakes and marshes - the rain also falls directly on these areas.

As the acidity of a lake increases, the water becomes clearer and the numbers of fish and other water animals decline. Some species of plant and animal are better able to survive in acidic water than others. Freshwater shrimps, snails, mussels are the most quickly affected by acidification followed by fish such as minnows, salmon and roach. The roe and fry (eggs and young) of the fish are the worst affected, the acidity of the water can cause deformity in young fish and can prevent eggs from hatching properly.

Lakes, rivers and marshes each have their own fragile ecosystem with many different species of plants and animals all depending on one another to survive. If a species of fish disappears, the animals which feed on it will gradually disappear too. If the extinct fish used to feed on a particular species of large insect, that insect population will start to grow, this in turn will affect the smaller insects or plankton on which the larger insect feeds.

Buildings

Every type of material will become eroded sooner or later by the effects of the climate. Water, wind, ice and snow all help in the erosion process but unfortunately; acid rain can help to make this natural process even quicker. Statues, buildings, vehicles, pipes and cables can all suffer. The worst affected are things made from limestone or sandstone as these types of rock are particularly susceptible and can be affected by air pollution in gaseous form as well as by acid rain.

Origin of Acid Rain

Until relatively recently air pollution has been seen as a local issue. It was in southern Scandinavia in the late 1950's that the problems of acid rain were first observed and it was then that people began to realise that the origins of this pollution were far away in Britain and Northern Europe. One early answer to industrial air pollution was to build very tall chimneys. Unfortunately all this does is push the polluting gases up into the clouds allowing emissions to float away on the wind. The wind carries the pollution many hundreds of miles away where it eventually falls as acid rain. In this way Britain has contributed at least 16% of the acid deposition in Norway. Over ninety percent of Norway's acid pollution comes from other countries. The worst European polluters are Germany, UK, Poland and Spain, each of them producing over a million tons of sulphur emissions in 1994. Governments are now beginning to admit that acid rain is a serious environmental problem and many countries are now taking steps to reduce the amount of sulphur and nitrogen emissions.

Measures

Reduce emissions:

- Burning fossil fuels is still one of the cheapest ways to produce electricity so people are now researching new ways to burn fuel which don't produce so much pollution.

- Governments need to spend more money on pollution control even if it does mean an increase in the price of electricity.
- Sulphur can also be 'washed' out of smoke by spraying a mixture of water and powdered limestone into the smokestack.
- Cars are now fitted with catalytic converters which remove three dangerous chemicals from exhaust gases.

Alternative Sources of Energy

- Governments need to invest in researching different ways to produce energy.
- Two other sources that are currently used are hydroelectric and nuclear power. These are 'clean' as far as acid rain goes but what other impact do they have on our environment?
- Other sources could be solar energy or windmills but how reliable would these be in places where it is not very windy or sunny?
- All energy sources have different benefits and costs and all these have to be weighed up before any government decides which of them it is going to use.

Conserving Resources

- Greater subsidies of public transport by the government to encourage people to use public transport rather than always travelling by car.
- Every individual can make an effort to save energy by switching off lights when they are not being used and using energy-saving appliances - when less electricity is being used, pollution from power plants decreases.
- Walking, cycling and sharing cars all reduce the pollution from vehicles.

Restoring the Damage done by Acid Rain

Lakes and rivers can have powdered limestone added to them to neutralise the water - this is called "liming". Liming, however, is expensive and its effects are only temporary - it needs to be continued until the acid rain stops. The people of Norway and Sweden have successfully used liming to help restore lakes and streams in their countries. A major liming programme is currently taking place in Wales.

Thunderstorm

Thunderstorm is a violent, short-lived weather disturbance that is almost always associated with lightning, thunder, dense clouds, heavy rain or hail, and strong, gusty winds. Thunderstorms arise when layers of warm, moist air rise in a large, swift updraft to cooler regions of the atmosphere. There the moisture contained in the updraft condenses to form towering cumulonimbus clouds and, eventually, precipitation. Columns of cooled air then sink earthward, striking the ground with strong downdrafts and horizontal winds. At the same time, electrical charges accumulate on cloud particles (water droplets and ice). Lightning discharges occur when the accumulated electric charge becomes sufficiently large. Lightning heats the air it passes through so intensely and quickly that shock waves are produced; these shock waves are heard as claps and rolls of thunder. On occasion, severe thunderstorms are accompanied by swirling vortices of air that become concentrated and powerful enough to form tornadoes.

Thunderstorms are known to occur in almost every region of the world, though they are rare in Polar Regions and infrequent at latitudes higher than 50 °N and 50 °S. The temperate and tropical regions of the world, therefore, are the most prone to thunderstorms. In the United States the areas of maximum thunderstorm activity are the Florida peninsula (more than 90 thunderstorm days per year), the Gulf Coast (70–80 days per year), and the mountains of New Mexico (50–60 days per year). Central Europe and Asia average 20 to 60 thunderstorm days per year. It has been estimated that at any one moment there are approximately 1,800 thunderstorms in progress throughout the world.

Thunderstorm Formation and Structure

Vertical Atmospheric Motion

Most brief but violent disturbances in Earth's wind systems involve large areas of ascending and descending air. Thunderstorms are no exception to this pattern. In technical terms, a thunderstorm is said to develop when the atmosphere becomes "unstable to vertical motion." Such instability can arise whenever relatively warm; light air is overlain by cooler, heavier air. Under such conditions the cooler air tends to sink, displacing the warmer air upward. If a sufficiently large volume of air rises, an updraft (a strong current of rising air) will be produced. If the updraft is moist, the water will condense and form clouds; condensation in turn will release latent heat energy, further fueling upward air motion and increasing the instability.

Once upward air motions are initiated in an unstable atmosphere, rising parcels of warm air accelerate as they rise through their cooler surroundings because they have a lower density and are more buoyant. This motion can set up a pattern of convection wherein heat and moisture are transported upward and cooler and drier air is transported downward. Areas of the atmosphere where vertical motion is relatively strong are called cells, and when they carry air to the upper troposphere (the lowest layer of the atmosphere), they are called deep cells. Thunderstorms develop when deep cells of moist convection become organized and merge, and then produce precipitation and ultimately lightning and thunder.

Upward motions can be initiated in a variety of ways in the atmosphere. A common mechanism is by the heating of a land surface and the adjacent layers of air by sunlight. If surface heating is sufficient, the temperatures of the lowest layers of air will rise faster than those of layers aloft, and the air will become unstable. The ability of the ground to heat up quickly is why most thunderstorms form over land rather than oceans. Instability can also occur when layers of cool air are warmed from below after they move over a warm ocean surface or over layers of warm air. Mountains, too, can trigger upward atmospheric motion by acting as topographic barriers that force winds to rise. Mountains also act as high-level sources of heat and instability when their surfaces are heated by the Sun.

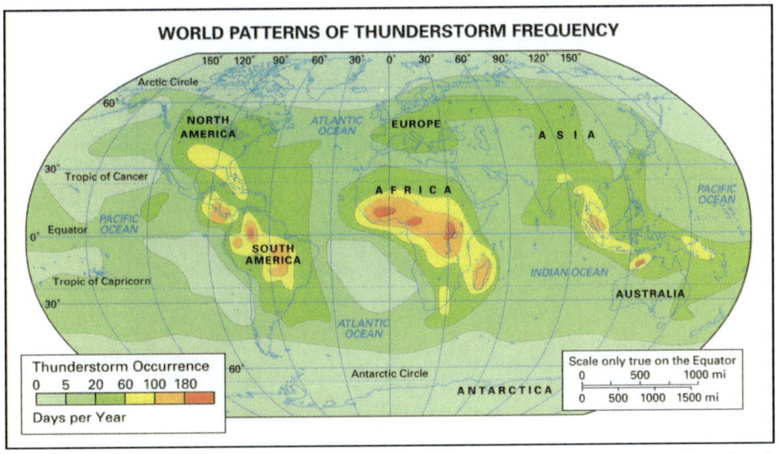

World patterns of thunderstorm frequency: Thunderstorms occur most often in the tropical latitudes over land, where the air is most likely to heat quickly and form strong updrafts.

The huge clouds associated with thunderstorms typically start as isolated cumulus clouds that develop vertically into domes and towers. If there is enough instability and moisture and the background winds are favourable, the heat released by condensation will further enhance the buoyancy of the rising air mass. The cumulus clouds will grow and merge with other cells to form a cumulus congestus cloud extending even higher into the atmosphere (6,000 metres [20,000 feet] or more above the surface). Ultimately, a cumulonimbus cloud will form, with its characteristic anvil-shaped top, billowing sides, and dark base. Cumulonimbus clouds typically produce large amounts of precipitation.

Types of Thunderstorms

At one time, thunderstorms were classified according to where they occurred—for example, as local, frontal, or orographic (mountain-initiated) thunderstorms. Today it is more common to classify storms according to the characteristics of the storms themselves, and such characteristics depend largely on the meteorological environment in which the storms develop. The United States National Weather Service has defined a severe thunderstorm as any storm that produces a tornado, winds greater than 26 metres per second (94 km [58 miles] per hour), or hail with a diameter greater than 1.9 cm (0.75 inch).

Isolated Thunderstorms

Isolated thunderstorms tend to occur where there are light winds that do not change dramatically with height and where there is abundant moisture at low and middle levels of the atmosphere—that is, from near the surface of the ground up to around 10,000 metres (33,000 feet) in altitude.

These storms are sometimes called air-mass or local thunderstorms. They are mostly vertical in structure, are relatively short-lived, and usually do not produce violent weather at the ground. Aircraft and radar measurements show that such storms are composed of one or more convective cells, each of which goes through a well-defined life cycle. Early in the development of a cell, the air motion is mostly upward, not as a steady, uniform stream but as one that is composed of a series of rising eddies. Cloud and precipitation particles form and grow as the cell grows. When the accumulated load of water and ice becomes excessive, a downdraft starts. The downward motion is enhanced when the cloud particles evaporate and cool the air—almost the reverse of the processes in an updraft. At maturity, the cell contains both updrafts and downdrafts in close proximity. In its later stages, the downdraft spreads throughout the cell and diminishes in intensity as precipitation falls from the cloud. Isolated thunderstorms contain one or more convective cells in different stages of evolution. Frequently, the downdrafts and associated outflows from a storm trigger new convective cells nearby, resulting in the formation of a multiple-cell thunderstorm.

Rain and lightning during a thunderstorm in Arizona.

Solar heating is an important factor in triggering local, isolated thunderstorms. Most such storms occur in the late afternoon and early evening, when surface temperatures are highest.

Multiple-cell Thunderstorms and Mesoscale Convective Systems

Violent weather at the ground is usually produced by organized multiple-cell storms, squall lines, or a supercell. All of these tend to be associated with a mesoscale disturbance (a weather system of intermediate size, that is, 10 to 1,000 km [6 to 600 miles] in horizontal extent). Multiple-cell storms have several updrafts and downdrafts in close proximity to one another. They occur in clusters of cells in various stages of development moving together as a group. Within the cluster one cell dominates for a time before weakening, and then another cell repeats the cycle. In squall lines, thunderstorms form in an organized line and create a single, continuous gust front (the leading edge of a storm's outflow from its downdraft). Supercell storms have one intense updraft and downdraft.

Sometimes the development of a mesoscale weather disturbance causes thunderstorms to develop over a region hundreds of kilometres in diameter. Examples of such disturbances include frontal wave cyclones (low-pressure systems that develop from a wave on a front separating warm and cool air masses) and low-pressure troughs at upper levels of the atmosphere. The resulting pattern of storms is called a mesoscale convective system (MCS). Severe multiple-cell thunderstorms and supercell storms are frequently associated with MCSs. Precipitation produced by these systems typically includes rainfall from convective clouds and from stratiform clouds (cloud layers with a large horizontal extent). Stratiform precipitation is primarily due to the remnants of older cells with a relatively low vertical velocity—that is, with limited convection occurring.

Lightning: cloud-to-ground: Cloud-to-ground lightning discharge in a field from a cumulonimbus cloud.

Thunderstorms can be triggered by a cold front that moves into moist, unstable air. Sometimes squall lines develop in the warm air mass tens to hundreds of kilometres ahead of a cold front. The tendency of prefrontal storms to be more or less aligned parallel to the front indicates that they are initiated by atmospheric disturbances caused by the front.

In the central United States, severe thunderstorms commonly occur in the springtime, when cool westerly winds at middle levels (3,000 to 10,000 metres [10,000 to 33,000 feet] in altitude) move over warm and moist surface air flowing northward from the Gulf of Mexico. The resulting broad region of instability produces MCSs that persist for many hours or even days.

In the tropics, the northeast trade winds meet the southeast trades near the Equator, and the resulting intertropical convergence zone (ITCZ) is characterized by air that is both moist and unstable. Thunderstorms and MCSs appear in great abundance in the ITCZ; they play an important role in the transport of heat to upper levels of the atmosphere and to higher latitudes.

Supercell Storms

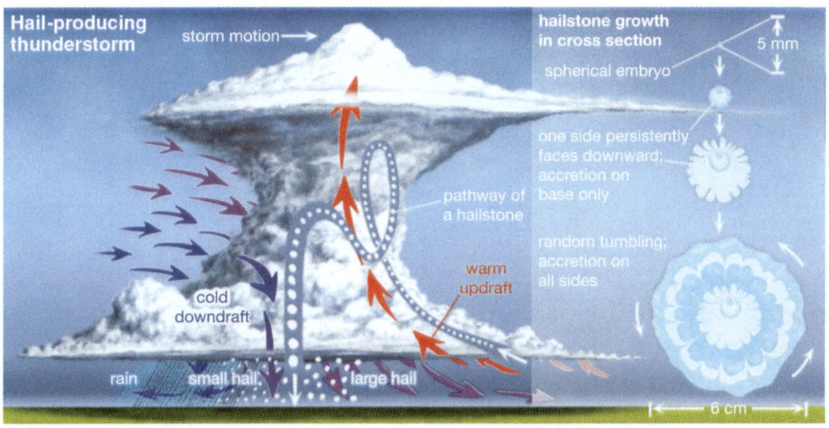

Hail-producing thunderstorm: (Left) A hailstone can travel through much of the height of the storm during its development and may make multiple vertical loops. (Right) Most hailstones are formed by accretion around a nucleus (spherical embryo). Peculiarly shaped hailstones are generally the product of multiple stones fusing together.

When environmental winds are favourable, the updraft and downdraft of a storm become organized and twist around and reinforce each other. The result is a long-lived supercell storm. These storms are the most intense type of thunderstorm. In the central United States, supercells typically have a broad, intense updraft that enters from the southeast and brings moist surface air into the storm. The updraft rises, rotates counter clockwise, and exits to the east, forming an anvil. Updraft speeds in supercell storms can exceed 40 metres (130 feet) per second and are capable of suspending hailstones as large as grapefruit. Supercells can last two to six hours. They are the most likely storm to produce spectacular wind and hail damage as well as powerful tornadoes.

Physical Characteristics of Thunderstorms

Aircraft and radar measurements show that a single thunderstorm cell extends to an altitude of 8,000 to 10,000 metres (26,000 to 33,000 feet) and lasts about 30 minutes. An isolated storm usually contains several cells in different stages of evolution and lasts about an hour. A large storm can be many tens of kilometres in diameter with a top that extends to altitudes above 18 km (10 miles), and its duration can be many hours.

Updrafts and Downdrafts

The updrafts and downdrafts in isolated thunderstorms are typically between about 0.5 and 2.5 km (0.3 and 1.6 miles) in diameter at altitudes of 3 to 8 km (1.9 to 5 miles). The updraft diameter may occasionally exceed 4 km (2.5 miles). Closer to the ground, drafts tend to have a larger diameter and lower speeds than do drafts higher in the cloud. Updraft speeds typically peak in the range of 5 to 10 metres (16 to 33 feet) per second, and speeds exceeding 20 metres (66 feet) per second are common in the upper parts of large storms. Airplanes flying through large storms at altitudes of about 10,000 metres (33,000 feet) have measured updrafts exceeding 30 metres (98 feet) per second. The strongest updrafts occur in organized storms that are many tens of kilometres in diameter, and lines or zones of such storms can extend for hundreds of kilometres.

Downbursts

Sometimes thunderstorms will produce intense downdrafts that create damaging winds on the ground. These downdrafts are referred to as macrobursts or microbursts, depending on their size. A macroburst is more than 4 km (2.5 miles) in diameter and can produce winds as high as 60 metres per second, or 215 km per hour (200 feet per second, or 135 miles per hour). A microburst is smaller in dimension but produces winds as high as 75 metres per second, or 270 km per hour (250 feet per second, or 170 miles per hour) on the ground. When the parent storm forms in a wet, humid environment, the microburst will be accompanied by intense rainfall at the ground. If the storm forms in a dry environment, however, the precipitation may evaporate before it reaches the ground (such precipitation is referred to as virga), and the microburst will be dry.

Thunderstorm microburst.

Figure indicates (Left) the air that forms the microburst is initially "dammed" aloft by the strength of the storm's updraft then cascades downward in a high-velocity, narrow column (less than 4 km, or 2.5 miles, in diameter). (Right, inset) Microbursts are very dangerous to aircraft and can create great damage on the ground. In the absence of observers, microburst damage can often be distinguished from that of a tornado by the presence of a "starburst" pattern of destruction radiating from a central point.

Downbursts are a serious hazard to aircraft, especially during take-offs and landings, because they produce large and abrupt changes in the wind speed and direction near the ground.

Vertical Extent

In general, an active cloud will rise until it loses its buoyancy. A loss of buoyancy is caused by precipitation loading when the water content of the cloud becomes heavy enough, or by the entrainment of cool, dry air, or by a combination of these processes. Growth can also be stopped by a capping inversion, that is, a region of the atmosphere where the air temperature decreases slowly, is constant, or increases with height.

Thunderstorms typically reach altitudes above 10,000 metres (33,000 feet) and sometimes more than 20,000 metres (66,000 feet). When the instability is high, the atmosphere moist, and winds favourable, thunderstorms can extend to the tropopause, that is, the boundary between the troposphere and the stratosphere. The tropopause is characterized by air temperatures that are nearly constant or increasing with height, and it is a region of great stability. Occasionally the momentum of an updraft carries it into the stratosphere, but after a short distance the air in the top of the updraft becomes cooler and heavier than the surrounding air, and the overshoot ceases. The height of the tropopause varies with both latitude and season. It ranges from about 10,000 to 15,000 metres (33,000 to 50,000 feet) and is higher near the Equator.

When a cumulonimbus cloud reaches a capping inversion or the tropopause, it spreads outward and forms the anvil cloud so characteristic of most thunderstorms. The winds at anvil altitudes typically carry cloud material downwind, and sometimes there are weak cells of convection embedded in the anvil.

Turbulence

An airplane flying through a thunderstorm is commonly buffeted upward and downward and from side to side by turbulent drafts in a storm. Atmospheric turbulence causes discomfort for the crew and passengers and also subjects the aircraft to undesirable stresses.

Turbulence can be quantified in various ways, but frequently a g unit, equal to the acceleration of gravity (9.8 metres per second squared, or 32.2 feet per second squared), is used. A gust of 1 g will cause severe aircraft turbulence. In the upper part of violent thunderstorms, vertical accelerations of about 3 g have been reported.

Movement of Thunderstorms

The motion of a thunderstorm across the land is determined primarily by the interactions of its updrafts and downdrafts with steering winds in the middle layers of the atmosphere in which the storm develops. The speed of isolated storms is typically about 20 km (12 miles) per hour, but some storms move much faster. In extreme circumstances, a supercell storm may move 65 to 80 km (about 40 to 50 miles) per hour. Most storms continually evolve and have new cells developing while old ones dissipate. When winds are light, an individual cell may move very little, less than two kilometres, during its lifetime; however, in a larger storm, new cells triggered by the outflow from downdrafts can give the appearance of rapid motion. In large, multicell storms, the new cells tend to form to the right of the steering winds in the Northern Hemisphere and to the left in the Southern Hemisphere.

Energy

The energy that drives thunderstorms comes primarily from the latent heat that is released when water vapour condenses to form cloud drops. For every gram of water that is condensed, about 600 calories of heat are released to the atmosphere. When water drops freeze in the upper parts of the cloud, another 80 calories per gram are released. The release of latent heat energy in an updraft is converted, at least in part, to the kinetic energy of the air motions. A rough estimate of the total energy in a thunderstorm can be made from the total quantity of water that is precipitated by the cloud. In a typical case, this energy is about 107 kilowatt-hours, roughly equivalent of a 20-kiloton nuclear explosion (though it is released over a broader area and in a longer span of time). A large, multicell storm can easily be 10 to 100 times more energetic.

Weather under Thunderstorms

Downdrafts and Gust Fronts

Thunderstorm downdrafts originate at altitudes where the air temperature is cooler than at ground level, and they are kept cool even as they sink to warmer levels by the evaporation of water and melting of ice particles. Not only is the sinking air denser than its surroundings, but it carries a horizontal momentum that is different from the surrounding air. If the descending air originated at a height of 10,000 metres (33,000 feet), for example, it might reach the ground with a horizontal velocity much higher than the wind at the ground. When such air hits the ground, it usually moves outward ahead of the storm at a higher speed than the storm itself. This is why an observer on the ground watching a thunderstorm approach can often feel a gust of cool air before the storm passes overhead. The outspreading downdraft air forms a pool some 500 to 2,000 metres (about 1,600 to 6,500 feet) deep, and often there is a distinct boundary between the cool air and the warm, humid air in which the storm developed. The passage of such a gust front is easily recognized as the wind speed increases and the air temperature suddenly drops. Over a five-minute period, a cooling of more than 5 °C (9 °F) is not unusual, and cooling twice as great is not unknown.

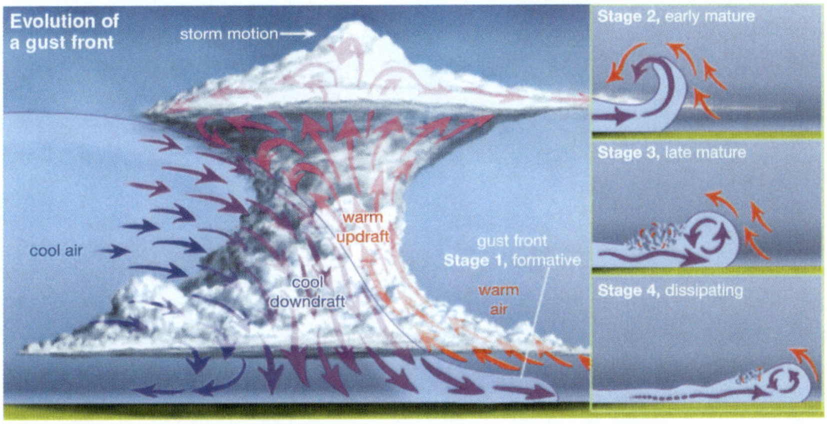

Evolution of a gust front.

Figure indicates (Left) during a thunderstorm a large column of cold air, originating high in the thundercloud, can descend rapidly to form a gust front. (Right, inset) Fed by the main downdraft, the gust front flows in a turbulent layer along the ground and can extend far from the main body of the storm. A gust front is often felt by observers as a sudden cool wind arriving well in advance of a storm.

Rainfall

In extreme circumstances, the gust front produced by a downburst may reach 50 metres (about 160 feet) per second or more and do extensive damage to property and vegetation. Severe winds occur most often when organized lines of thunderstorms develop in an environment where the middle-level winds are very strong. Under such conditions, people might think the winds were caused by a tornado. If a funnel cloud is not observed, the character of the wind damage can indicate the source. Tornadoes blow debris in a tight circular pattern, whereas the air from a thunderstorm outflow pushes it mostly in one direction.

By the time the cool air arrives, rain usually is reaching the surface. Sometimes all the raindrops evaporate while falling, and the result is a dry thunderstorm. At the other extreme, severe multiple-cell and supercell storms can produce torrential rain and hail and cause flash floods.

In small thunderstorms, peak five-minute rainfall rates can exceed 120 mm (4.7 inches) per hour, but most rainfalls are about one-tenth this amount. The average thunderstorm produces about 2,000 metric tons (220,000 short tons) of rain, but large storms can produce 10 times more rainfall. Large, organized storms that are associated with mesoscale convective systems can generate 10^{10} to 10^{12} kg of rainfall.

Thunderstorm Electrification

Within a single thunderstorm, there are updrafts and downdrafts and a variety of cloud particles and precipitation. Measurements show that thunderclouds in different geographic locations tend to produce an excess negative charge at altitudes where the ambient air temperature is between about −5 and −15 °C (23 to 5 °F). Positive charge accumulates at both higher and lower altitudes. The result is a division of charge across space that creates a high electric field and the possibility of significant electrical activity.

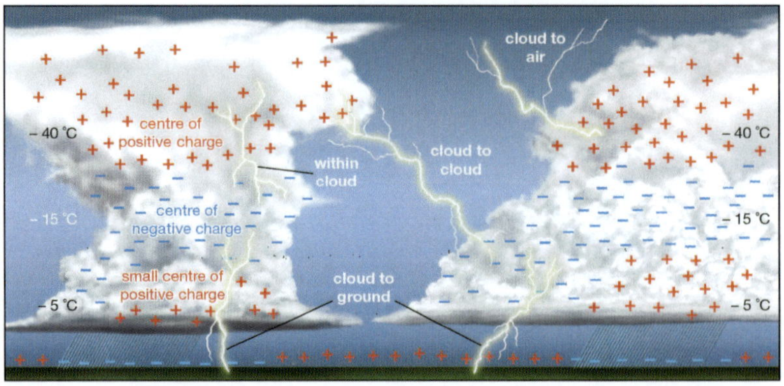

Electrical charge distribution in a thunderstorm.

Figure indicates when the electrical charges become sufficiently separated in a thundercloud, with some regions acquiring a negative charge and others a positive, a discharge of lightning becomes likely. About one-third of lightning flashes travel from the cloud to the ground; most of these originate in negatively charged regions of the cloud.

Many mechanisms have been proposed to explain the overall electrical structure of a thunderstorm, and cloud electrification is an active area of research. A leading hypothesis is that if the larger and heavier cloud particles charge preferentially with a negative polarity, and the smaller and lighter particles acquire a positive polarity, then the separation between positive and negative regions occurs simply because the larger particles fall faster than the lighter cloud constituents. Such a mechanism is generally consistent with laboratory studies that show electrical charging when soft hail, or graupel particles (porous amalgamations of frozen water droplets), collide with ice crystals in the presence of super cooled water droplets. The amount and polarity of the graupel charges depend on the ambient air temperature and on the liquid water content of the cloud, as well as on the ice crystal size, the velocity of the collision, and other factors. Other mechanisms of electrification are also possible.

Lightning Occurrence

Lightning flash striking a tree at a distance of 60 metres (200 feet) from the camera.

When the accumulated electric charges in a thunderstorm become sufficiently large, lightning discharges take place between opposite charge regions, between charged regions and the ground, or from a charged region to the neutral atmosphere. In a typical thunderstorm, roughly two-thirds of all discharges occur within the cloud, from cloud to cloud, or from cloud to air. The rest are between the cloud and ground.

In recent years it has become clear that lightning can be artificially initiated, or triggered, in clouds that would not normally produce natural lightning discharges. Lightning can be triggered by a mountain or a tall structure when a thunderstorm is overhead and there is a high electric field in the vicinity or when an aircraft or large rocket flies into a high-field environment.

Lightning discharge triggered by the presence of a tall tower atop Mount San Salvatore, near Lugano, Switzerland.

Global Lightning Distribution

Data from Earth-orbiting satellites show that, on average, about 80 percent of lightning flashes occur over land and 20 percent over the oceans. The frequency of lightning over land tends to peak in the mid-afternoon between 3:00 and 6:00 PM local time. Seasonal trends in the distribution of lightning are the result of temperature changes at the Earth's surface.

Tropical air masses commonly produce thunderstorms and lightning. Thunderstorm development requires moist, unstable air masses typical of those in tropical areas. In this region the Sun's rays are nearly vertical, allowing more energy to reach and warm the lowest layers of the atmosphere. Abundant moisture is added when the warm air moves over the ocean and becomes humidified by evaporation from the underlying water surface. Thunderstorm development is then initiated by upward movement of air, due to, for example, changes in air pressure or the topography of the land. The average number of days with audible thunder exceeds 100 per year over land areas within 10 degrees latitude north and south of the Equator. In some regions of equatorial Africa and South America there are more than 180 thunder days in an average year.

At higher latitudes, thunderstorm frequency depends on the character of the topography and how often moist, tropical air invades the region, which happens most often in the spring and summer. Maximum thunderstorm activity in the Northern and Southern Hemispheres is offset by approximately six months, with most Northern Hemisphere thunderstorms occurring between May and September and in the Southern Hemisphere between November and March.

Thunderstorms are a common feature of the summer monsoons in many parts of the world, especially southern Asia. As solar radiation warms the Indian subcontinent, an ocean-to-land air current is established and moist, unstable air from the Indian Ocean is carried inland. When this air is forced to rise by the steep slopes of the Himalayas, intense thunderstorms and rain showers are produced in great abundance.

In regions poleward of about 60 degrees latitude thunderstorms are rare to non-existent. In these regions the air near the surface is cold and the atmosphere is generally stable. There are also few thunderstorms in regions that are dominated by semi-permanent high-pressure centres, such as southern California. In these regions air from higher altitudes is descending and warming, which lowers the relative humidity and causes stable stratification of the lower atmosphere. As a result, thunderstorm development is inhibited.

Lightning Distribution in the United States

Every year, most of the United States experiences at least two cloud-to-ground strikes per square kilometre (about five per square mile). Most of the interior of the country east of the Rocky Mountains has four or more strikes per square kilometre (about 10 discharges per square mile). Summer thunderstorms are frequent in northern Mexico and the states of Arizona, New Mexico, and Colorado when warm, humid air is forced to rise by mountainous terrain.

Maximum flash densities are found along the Gulf Coast and Florida peninsula, where over a year's time, values exceeding 10 strikes per square kilometre (25 strikes per square mile) have been measured. More than 20 million cloud-to-ground flashes strike the United States annually, and lightning is clearly among the country's most severe weather hazards.

Cloud-to-ground Lightning

Initial Stroke

A typical flash of cloud-to-ground lightning is initiated by electrical breakdown between the small positive charge region near the base of the cloud and the negative charge region in the middle of the cloud. The preliminary breakdown creates channels of air that have undergone partial ionization—the conversion of neutral atoms and molecules to electrically charged ones.

On timescales measured in fractions of a second, high-speed cameras can record luminous events in the flash. Initially, a faint luminous process descends in a downward-branching pattern in regular distinct steps, typically 30 metres (100 feet) in length, though they can range from 10 to 100 metres (33 to 330 feet). The time interval between steps ranges from 10 to 50 microseconds (millionths of a second). Carrying currents on the order of hundreds to thousands of amperes, the stepped leader propagates toward the ground at an average velocity of 1.5 × 105 metres per second, or about one two-thousandth the speed of light. It is called a stepped leader because of its downward-moving "stepped" pulses of luminosity. Diameter estimates for the stepped leader range from a few centimetres to a few metres. The current-carrying core has a diameter on the order of 1 or 2 cm (0.4 or 0.8 inch), and photographic measurements indicate that a corona sheath of electric charge with a diameter of 1 to 10 metres (3 to 33 feet) surrounds the core.

Return Stroke

As the stepped leader nears the ground, approximately five coulombs of charge have been deposited along the channel, inducing an opposite charge on the ground and increasing the electric field between the leader and the point to be struck. An upward discharge starts at the ground, church steeple, house, or other object, and rises to meet the stepped leader about 15 to 50 metres (50 to 160 feet) above the surface. At this moment of junction the cloud is short-circuited to the ground and a highly luminous return stroke of high current occurs. It is this return stroke, rather than the stepped leader, that is perceived as lightning because it is so much brighter and follows so quickly after the stepped leader. Portions of the stepped leader that have not reached the ground become the branches of the return stroke, and charge on the branches flows into the main channel. The five coulombs of charge typically deposited along the stepped leader flow to ground in a few hundred microseconds and produce peak currents that are usually on the order of 30,000 amperes but may range from a few thousand to over 200,000 amperes. Peak temperatures in the channel are on the order of 30,000 °C (50,000 °F), about five times hotter than the surface of the Sun. Because the junction process occurs near the ground, the time to peak current measured at the ground is typically only a few microseconds. As the leader charge avalanches toward the ground, the return stroke luminosity propagates toward the cloud base at an average speed of 5×10^7 to 2×10^8 metres per second, or approximately one-third the speed of light, and the high-current-carrying core expands to a diameter of a few centimetres. Laboratory experiments suggest that when pressure equilibrium is attained between the return stroke and the surrounding air, the channel approximates a high-current arc characterized by a current density of 1,000 amperes per square centimetre.

Cloud-to-ground lightning discharge showing a bright main channel and secondary branches.

Subsequent Return Strokes

In the rapid passage from ground to cloud, the luminous return stroke is observed to pause at points where large branches join the main channel, and the channel is observed to brighten as charge from the branch flows into the channel. The stroke then continues its upward propagation, reaching the level of the atmosphere where the temperature is 0 °C (typically at an altitude of 5 km [3 miles] above sea level) in approximately 100 microseconds; the downward-propagating stepped leader traverses the same distance in about 30 milliseconds (thousandths of a second). There is then a pause for tens of milliseconds, and the channel cools to a few thousand degrees Celsius. If a second stroke occurs, it begins with the appearance of a dart of light, perhaps 30 to 50 metres (100 to 160 feet) in length, propagating down the channel of the previous return stroke. The dart leader moves downward at a speed of 2×10^6 metres per second (about one one-hundredth the speed of light) and carries a current of the order of 1,000 amperes toward the ground. Once again, when the leader effectively short-circuits a charge centre in the cloud to the ground, another return stroke occurs. After the first stroke, the dart leader may follow the lightning channel only partway before taking a new path to the ground. This gives rise to the common forked appearance of lightning as it strikes the ground.

This sequence of dart leader-return stroke typically occurs three to four times, although a flash to the ground that had 26 strokes and lasted two seconds has been reported. When a flash does have more than one stroke, the subsequent return strokes draw charge from different regions of the parent thunderstorm. Multiple strokes of lightning appear to flicker because the human eye is just capable of resolving the time interval between them.

Dissipation of Energy

During the return-stroke stage, approximately 10^5 joules of energy per metre are dissipated within the lightning channel. This energy is divided among the dissociation, ionization, excitation, and kinetic energy of the particles, the energy of expansion of the channel, and radiation. Spectroscopic measurements reveal that the air molecules, principally those of nitrogen, oxygen, and water, are split into their respective atoms and that on the average one electron is removed from each atom. The conversion from neutral air molecules to a completely ionized plasma occurs in a few microseconds.

Thunder

When the stroke plasma is created, its temperature is at least 30,000 °C (50,000 °F), and the pressure is greater than 1,000 kilopascals (10 atmospheres). The channel pressure greatly exceeds the ambient (surrounding) pressure, and the return-stroke channel expands at a supersonic rate. The resultant shock wave decays rapidly with distance and is eventually heard as thunder once it slows to the speed of sound. Because it is estimated that only 1 percent of the input energy is stored in the particles and less than 1 percent is emitted as radiation in the visible and infrared region (4,000 to 11,000 angstroms [Å], where Å = 10^{-10} metre), it is probable that most of the energy dissipated goes into the energy of channel expansion, a process requiring no more than 10 to 20 microseconds.

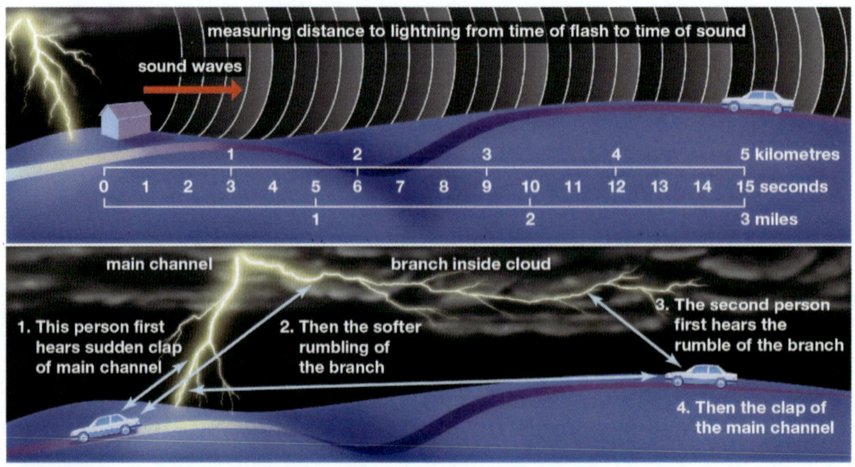

In figure, (top) as shown in the chart, the elapsed time between seeing a flash of lightning and hearing the thunder is roughly three seconds for each kilometre, or five seconds for each mile. (Bottom) An observer's relative distance from the main lightning channel and its secondary branches determines whether thunder is heard to start with a sudden clap or a softer rumbling.

Since light travels at about 300,000 km (186,000 miles) per second and the speed of sound is only about 0.33 km (0.2 mile) per second, the light from a discharge will always be seen before the sound arrives at an observer. The time delay between the bright flash of light and the arrival of the associated thunder can often be used to estimate the distance to a discharge. Every three seconds correspond to one kilometre, and every five seconds correspond to one mile.

The total thunder waveform comes from the entire lightning channel and includes the effects of channel branching and tortuosity, sound propagation in the atmosphere, and acoustic reflections from the local topography. The result is a series of sounds that are variously described as peaks, claps, rolls, and rumbles. At distances of a few hundred metres, thunder begins with a sudden clap followed by a long rumble; at larger distances, it begins with a rumble.

Triggered Lightning

A small percentage of discharges between the cloud and ground are actually initiated at the ground and propagate upward to a charged region in the cloud. These discharges often are initiated (or triggered) by tall structures or by towers on hilltops. The upward branching of such discharges makes them visually distinguishable from their "right-side-up" counterparts, giving the impression of a cloud-to-ground lightning flash that is upside down.

Cloud-to-cloud and Intracloud Lightning

True cloud-to-cloud lightning is rare because most lightning flashes occur within a cloud. The first lightning flash in a thunderstorm is typically an intracloud discharge. When an intracloud discharge occurs, the cloud becomes luminous for approximately 0.2 to 0.5 second. The discharge is initiated by a leader that propagates between regions of opposite charge (or from a charged region to the neutral atmosphere). Luminosity is more or less continuous and has several pulses of higher luminosity of one-millisecond duration superimposed upon it. This situation suggests minor return strokes as the leader contacts pockets of opposite charge, but the similarity ends there. The total amount of the charge transfer is generally similar to the amount involved in a ground discharge: 10 coulombs, with a range from 0.3 to 100 coulombs. The mean velocity of propagation of intracloud lightning ranges from 10^4 to 10^7 metres per second. Electric currents associated with the luminous brightening are probably in the range of 1,000 to 4,000 amperes. Strikes to aircraft exhibit peak currents of only a few thousand amperes, about an order of magnitude less than currents in ground flashes—though sometimes the peak currents are large. Rise times to peak currents in cloud flashes are generally slower than those in return strokes. The amount of energy dissipated by intracloud flashes is unknown.

Lightning Damage

Most lightning strikes cause damage through the large current flowing in the return stroke or through the heat that is generated by this and the continuing current. The precise mechanisms whereby lightning currents cause damage are not completely understood, however. If lightning strikes a person, the stroke current can damage the central nervous system, heart, lungs, and other vital organs.

When a building or power line is struck by lightning or is exposed to the intense electromagnetic fields from a nearby flash, the currents and voltages that appear on the structure are determined both by the currents and fields in the discharge and by the electrical response of the object and its grounding system. For instance, if a lightning surge enters an unprotected residence by way of an electric power line, the voltages may be large enough to cause sparks in the house wiring or appliances. When such flashovers occur, they may short-circuit the alternating current power system, and the resulting power arc may start a fire. In such instances, the lightning does not start the fire directly, but it does cause a power fault (short circuit), and then the power currents do the damage. In the case of metals, large currents heat the surface at the air-arc interface and the interior by electron collisions with the metal lattice. If this heat is also great enough, the metal will melt or evaporate.

At least three properties of the return-stroke current can cause damage; these are the peak current, the maximum rate of change of the initial current, and the total amount of charge transferred. For objects that have resistive impedance, such as a ground rod or a long power line, the peak voltage during a strike is proportional to the peak current produced of the lightning stroke and the resistivity of the struck object. For example, if a 100,000 ampere peak current flows into a 10-ohm grounding system, 1 million volts will be produced. A common hazard associated with the large voltages produced by lightning strikes is the re-direction of some of the energy (that is, a flashover) from the original target to an adjacent object. Such secondary discharges, or side-flashes, often cause damage comparable to that of a direct strike, and they are one of the main hazards of standing under or near an isolated tree (or any other tall object) during a thunderstorm. Such large voltages frequently cause secondary discharges or side-flashes to radiate outward from the object that is struck to another object nearby. One form of a side-flash can even occur in the ground near the point of lightning attachment.

For objects that have an inductive electrical impedance, such as the wires in a home electrical system, the peak voltage will be proportional to the maximum rate of change of the lightning current and the inductance of the object. For example, one metre of straight copper wire has a self-inductance on the order of one microhenry. The peak rate of change in the lightning current in a return stroke is on the order of 100,000 amperes per microsecond; therefore, about 100,000 volts will appear across this length of conductor for the duration of the change, typically 100 nanoseconds (billionths of a second).

The heating and subsequent burn-through of metal sheets, as on a metal roof or tank, are to a first approximation proportional to the total charge injected into the metal at the air-arc interface. Generally, large charge transfers are produced by long-duration continuing currents that are in the range of 100 to 1,000 amperes, rather than by the peak currents, which have a relatively short duration. The heat produced by long continuing currents is frequently the cause of forest fires. A typical cloud-to-ground flash transfers 20 to 30 coulombs of charge to the ground, and extreme flashes transfer hundreds and occasionally thousands of coulombs.

Lightning Protection

The best personal protection against lightning is to be alert to the presence of a hazard and then to take common-sense precautions, such as staying inside a house or building or inside an automobile, where one is surrounded by (but not in contact with) metal. People are advised to stay away from outside doors and windows and not to be in contact with any electrical appliances, such as a telephone, or anything connected to the plumbing system. If caught outdoors, people are advised to avoid isolated trees or other objects that are preferred targets and to keep low so as to minimize both height and contact with the ground (that is, crouch but do not lie down). Swimming pools are not safe during a lightning storm because water is a good conductor of electricity, and hence being in the pool effectively greatly multiplies the area of one's "ground" contact.

The frequency with which lightning will directly strike a building in a particular region can be estimated from the building's size and the average number of strikes that occur in the region. If a building is struck whenever a stepped leader comes within 10 metres (33 feet) of the exterior of the building, then a building that is 12 metres (39 feet) wide and 16 metres (52 feet) long (an area of 192 square metres, or about 2,000 square feet) will have an effective strike zone of 32 metres by 36 metres (an area of 1,152 square metres, or 12,400 square feet). In a region where an average of three cloud-to-ground lightning strikes occur per square kilometre annually, such a building will experience an average of 0.0035 direct strike per year, or one strike about every 290 years (1,152 square metres × 3 flashes per square kilometre × 10^{-6} metres per square kilometre). In a region where there is an annual average of five strikes per square kilometre, the same building will experience an average of 0.0058 direct strikes per year, or one strike about every 174 years. These calculations indicate that, for the second example, an average of one of every 174 buildings of similar size will be directly struck by lightning in that region each year.

Structures may be protected from lightning by either channelling the current along the outside of the building and into the ground or by shielding the building against damage from transient currents and voltages caused by a strike. Many buildings constrain the path of lightning currents and voltages through use of lightning rods, or air terminals, and conductors that route the current down into a grounding system. When a lightning leader comes near the building, the lightning rod initiates a discharge that travels upward and connects with it, thus controlling the point of attachment of lightning to the building. A lightning rod functions only when a lightning strike in the immediate vicinity is already immanent and so does not attract significantly more lighting to the building. The down conductors and grounding system function to guide the current into the ground while minimizing damage to the structure. To minimize side-flashes, the grounding resistance should be kept as low as possible, and the geometry should be arranged so as to minimize surface breakdown. Overhead wires and grounded vertical cones may also be used to provide a cone-shaped area of lightning protection. Such systems are most efficient when their height is 30 metres (98 feet) or less.

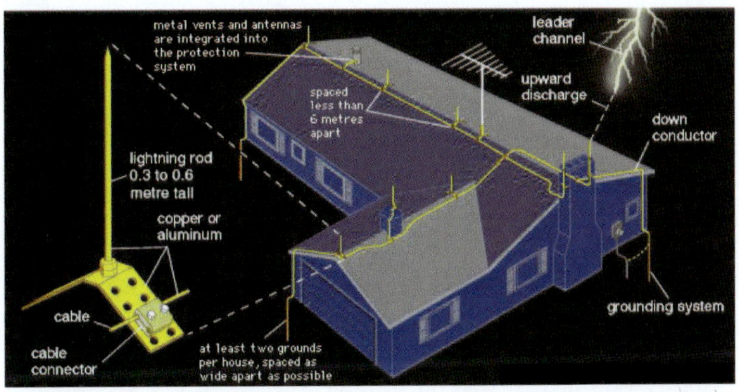

Lightning rod protection system for a residential building. The flow of electricity from a lightning strike is channeled harmlessly around the outside of the building and into the ground.

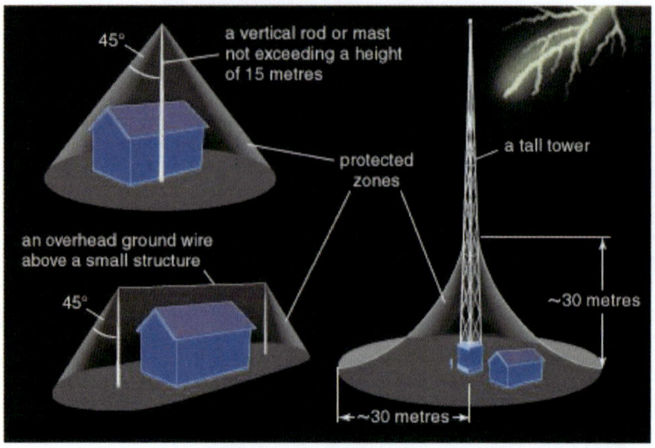

Lightning rod types.

Figure indicates vertical rods or masts up to 15 metres in height create lightning protection zones that extend in a 45° cone from the rod's tip. Connecting two rods with a wire extends the zone of protection. Towers taller than 30 metres provide protection for an area 30 metres high and 60 metres wide. The protected zone is in the shape of an inverted funnel with inward-curving sides. Towers between 15 and 30 metres high create protected zones of similar shape but with height and width equal to tower height.

Protection of the contents of a structure can be enhanced by using lightning arresters to reduce any transient currents and voltages that might be caused by the discharge and that might propagate into the structure as traveling waves on any electric power or telephone wires exposed to the outside environment. The most effective protection for complex structures is provided by topological shielding. This form of protection reduces amounts of voltage and power at each level of a system of successive nested shields. The partial metallic shields are isolated, and the inside surface of each is grounded to the outside surface of the next. Power surges along wires coming into the structure are deflected by arrestors, or transient protectors, to the outside surface of each shield as they travel through the series, and are thus incrementally attenuated.

Thundersnow

Thundersnow is a snowstorm accompanied by thunder and lightning. The phenomenon is rare, even in areas prone to snow. You're not likely to get thunder and lightning during a gentle snowfall. The weather needs to be seriously bad. Examples of storms with thundersnow include the bomb cyclone of 2018, Blizzard of 1978 (north-eastern United States), Winter Storm Niko (Massachusetts), and Winter Storm Grayson (New York).

Obviously, if it never gets cold enough to snow, thundersnow is out of the question. In any given year, an average of 6.4 events is reported worldwide. While thundersnow is uncommon under any circumstances, some locations have more favorable conditions than others:

- Great plains;
- Mountains;

- Coastlines;
- Lake-effect regions.

Areas reporting higher-than-average thundersnow events include the eastern side of the Great Lakes of the United States and Canada, the plains regions of the midwestern United States, the Great Salt Lake, Mount Everest, the Sea of Japan, Great Britain, and elevated regions of Jordan and Israel. Specific cities known to experience thundersnow include Bozeman, Montana; Halifax, Nova Scotia; and Jerusalem.

Thundersnow tends to occur late in the season, typically April or May in the Northern Hemisphere. The peak formation month is March. Coastal regions may experience sleet, hail, or freezing rain rather than snow.

Occurrence of Thundersnow

Thundersnow is rare because the conditions that produce snow tend to have a stabilizing effect on the atmosphere. In the winter, the surface and lower troposphere are cold and have low dew points. This means there is little moisture or convection to lead to lightning. Lightning superheats the air, while the rapid cooling produces the sound waves we call thunder.

Thunderstorms can form in winter, but they have different characteristics. A typical normal thunderstorm consists of tall, narrow clouds that rise from a warm updraft leading from the surface up to around 40,000 feet. Thundersnow usually forms when layers of flat snow clouds develop instability and experience dynamic lifting. Three causes lead to the instability.

- A normal thunderstorm at the edge of a warm or cold front can run into cold air, changing rain into freezing rain or snow.
- Synoptic forcing, such as might be seen in an extratropical cyclone, can lead to thundersnow. The flat snow clouds become bumpy or develop what are called "turrets." Turrets rise about the clouds, making the top layer unstable. Turbulence causes water molecules or ice crystals to gain or lose electrons. When the electrical charge difference between two bodies becomes large enough, lightning occurs.
- A cold air front passing over warmer water can produce thundersnow. This is the type of thundersnow most often seen near the Great Lakes or near and ocean.

Differences from a Normal Thunderstorm

The obvious difference between a typical thunderstorm and thundersnow is that a thunderstorm produces rain, while thundersnow is associated with snow. However, the thunder and lightning of thundersnow are different, too. Snow muffles sound, so thundersnow thunder sounds subdued and doesn't travel as far as it would in a clear or rainy sky. Normal thunder may be heard miles from its source, while thundersnow thunder tends to be restricted to a 2 to 3 mile (3.2 to 4.8 kilometer) radius from the lightning strike.

While thunder may be muted, lightning flashes are enhanced by reflective snow. Thundersnow lightning typically appears white or golden, rather than the usual blue or violet of thunderstorm lightning.

Thundersnow Hazards

The conditions that lead to thundersnow also lead to dangerously cold temperatures and poor visibility from blowing snow. Tropical force wind is possible. Thundersnow is most common with blizzards or severe winter storms.

Thundersnow lightning is more likely to have a positive electrical charge. The positive polarity lightning is more destructive than usual negative polarity lightning. Positive lightning may be up to ten times stronger than negative lightning, up to 300,000 amperes and one billion volts. Sometimes positive strikes occur over 25 miles away from the point of precipitation. Thundersnow lightning can cause a fire or damage a power line.

Precipitation

Precipitation is any form of water descending from the clouds to the ground, encompassing a broad range of forms, from rain to hail. Water vapor has a constant presence in the Earth's atmosphere—although to varying degrees—and when the right conditions exist, it comes together to form clouds and, eventually, precipitation. It is also one step in the hydrologic cycle where vast quantities of water cycle through the Earth's atmosphere, oceans, land, and biosphere over both short and long time scales. This cycling helps shape our weather and climate, supports plant growth, and makes life itself possible. Precipitation is an important element in this cycle, helping to keep the Earth in balance by redistributing water from the oceans and warm, tropical areas to the rest of the planet, providing valuable moisture.

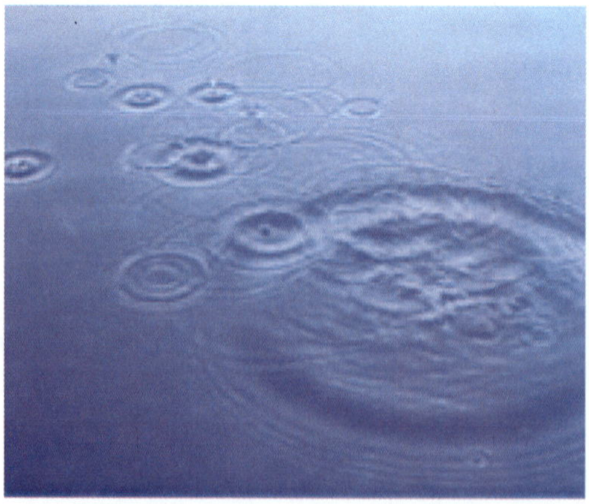

Liquid Precipitation

Rain, Drizzle and Freezing Rain

The differentiation between rain and drizzle is to a certain extent arbitrary. These two forms of liquid precipitation differ from each other only in the size of drops. The diameter of raindrops is usually 5 to 6 mm, whereas drizzle drops are smaller (between 0.2 and 0.5 mm) and their terminal velocities are between 70 and 200 cm per second. Drizzle falls mostly from low stratus clouds and is frequently accompanied with fog and poor visibility.

The diameter of raindrops is usually larger than 0.5 mm, but they only rarely reach 6 mm or more because larger raindrops are destroyed during falling. Small raindrops are of almost spherical shape, but bigger ones are flattened when falling, especially in the lower part of the cloud. The terminal velocities of rain drops range from two meters per second for the smallest to about 10 meters per second for the largest.

During heavy rains, raindrops are considerably bigger than in light rain. The largest drops of more than six millimeters in diameter appear only in heavy rains, especially at the start of a rain storm.

When raindrops pass through cold air layers (below 0 °C), they become super-cooled, and freezing rain or super-cooled drizzle occurs. Freezing rain falls in liquid form but freezes upon impact to form a coating of glaze on the ground and exposed objects. Often these frozen raindrops form a very slippery and almost transparent "glazed" film which is dangerous for both pedestrians and transport.

Ground Liquid Precipitation: Dew and Liquid Deposit

Dew is the smallest water drop formed during condensation on the Earth's surface, most frequently on grass during the warm period of a year. Dew arises mostly with clear and calm weather in the evening hours and at night when there is no fog. Dew develops when soil and vegetation (grass and leaves) cool to the temperature typical of the dew point.

Dew can produce a large amount of precipitation. For example, in England on flat ground, dew produces 10 to 30 millimeter of precipitation a year. In the mid-latitudes of Europe this amount is about 10 millimeter a year, and in southern Africa 40 millimeter a year. In warm, humid tropical regions, dew is so abundant that it can drip from roofs of buildings and tree leaves.

In Northern Europe (Finland), abundant dew can produce a water layer of 0.1 to 0.2 mm each night. From the data measured by Divdevany's dew meter at a number of Finnish stations during May to October in 1967 and 1968, the dew amount is about 10 mm, or 3% of precipitation for the warm period of the year. In Northern Finland, the dew in warm periods is as much as 1.5% of total precipitation for the same period.

There is a relationship between the dates of dew appearance, cloudiness, and air humidity. Dew mainly arises with a low cloudiness of 30 to 50% and relative humidity of 60 to 80%. In northern and north-western Russia, dew falls with air temperature above 7 to 8 °C. In southern regions of Russia, dew develops at temperatures above 5-6 °C. In some regions, where the temperature is above 0 °C throughout the year, dew develops continuously.

The liquid film is a water layer arising on cold, mostly vertical, surfaces in gloomy and windy weather. Liquid film arises from a cause different from dew. It is formed from warmer and humid air advection after prolonged cold weather. It occurs where water vapor condenses on cooled surfaces (e.g. buildings, walls, trees, and fences). The process is most apparent on the windward sides of objects: as if they are misted when covered with tiny drops. A liquid film also forms under artificial conditions, such as when warm and humid interior air condenses on the inside of cold windowpanes in a warm house.

Solid Precipitation

Solid precipitation is of more diverse forms than liquid ones. There include snow, snow and ice pellets, hail, etc. In addition, as stated above, there is a great variety of surface solid precipitation (surface hydrometeors).

Snow

Snow is solid atmospheric precipitation of different forms of ice crystals. Groups of ice crystals form snowflakes in the shape of six-cone stars, needles or multiple combinations. The size of snowflakes varies from one millimeter to a few centimeters depending on air temperature: the higher the temperature and weaker the wind, the larger the flakes. The largest flakes fall with heavy precipitation.

The diversity of snowflakes is endless. Mostly they are stars, columns, or their combinations. The velocity of snowflake fall is a function of their shape and air temperature, being in the range of 0.1 to 2 meter per second in motionless air. Figure shows the shapes of snowflakes according to the international classification.

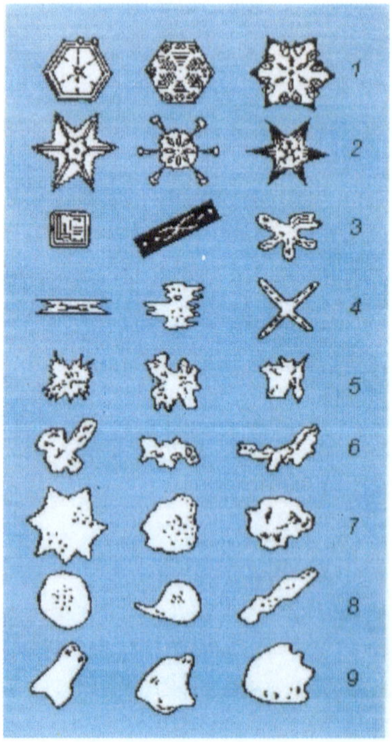

International classification of frozen precipitation. 1-Thin hexagonal plates; 2 -Dendritic crystals, 3-columns; 4-needles; 5-dendritic forms; 6-incorrect crystals; 7-snow pellets (graupel); 8-freezing rain; 9-hail.

Japanese researchers have shown a particular interest in snowflake shape. As long ago as the second half of the nineteenth century, Dai-Toshizura made a large number of delicate drawings of snowflakes of different shapes using a microscope. The artist reached a high level of accuracy with the original. His outstanding work remains unsurpassed.

Wilson Bentley in the US State of Vermont, took pictures of snowflakes for many years and developed a big collection of different shapes. His album containing 300 photos of snowflakes was published in 1931, the year of his death.

The Japanese researcher U. Nakaya (Hokkaido University) discovered when studying the process of developing snow crystals under man-made conditions (in a chamber) that ice crystals in the shape of stars develop only in the narrow temperature range of -14 to -17 °C. At temperature above -7 °C (from −3 °C to −5 °C) only ice needles are formed. Plates and columns are formed at temperatures form −10 °C to −22 °C.

Snow falls from clouds of different shapes, mostly strato-cumulus, high-stratus and cumulo-nimbus, during the cold period of the year. By their size snowflakes are classified as tiny (particles < 5 millimeter), small flakes (5 to 15 mm), or large flakes (> 15 mm). Snowflake fall velocity ranges from < 0.1 meter per second (slow flying) to > 0.8 meter per second. Plates and stars fall with velocity of 0.5 to 1 meter per second, needles and columns – a few decimeters per second, and snow and ice pellets at 1 to 2.7 meters per second.

The rate of snowfall is measured by precipitation amount in 1 mm water layers per unit of time, mostly per hour or 24 hours. Weak snow has a rate of less than 0.1 mm/hour, average snow – from 0.1 to 1 mm per hour, and heavy snow – more than 1mm per hour. Vision in dense snow can be reduced to 1 km or less. The length of snowfall period is usually inversely proportional to the rate of its fall. Dense snowfall rarely lasts more than 1 to 2 hours, and weak snow can last 24 hours or even longer.

At meteorological stations several types of snow are distinguished: snow with rain, rain with snow, drizzle, and heavy snow. The latter is sometimes called a snow shower. There is also such an event as a snow storm (or snow charge), as well as snow under clear sky.

Rain with snow usually falls at positive surface temperature and represents rain with a small amount of separate snowflakes. Snow with rain is a mixture of snowflakes and raindrops occurring at the temperatures close to 0 °C.

Overcasting snow falls mostly from stratocumulus or from high stratus clouds and can last for several hours continuously.

Pouring snow falls usually from cumulus-nimbus clouds at temperatures close to 0 °C. This kind of snow falls due to a cold front and unstable air mass. A snow storm is formed in an unstable stratified cold air mass passing over relatively warm underlying surface.

Sometimes colored snow occurs. The color is derived from mineral or organic admixtures. Colored snow is mostly of a brown or red shades due to dust settling on snow, or algae or bacteria reproducing on snow. Color snow mostly occurs in the high latitudes in spring when soil is partly bared and partly covered with snow.

Formation of Precipitation

The formation of precipitation may occur at temperatures above or below freezing. Precipitation that is formed in temperatures entirely above freezing is called warm precipitation; cold precipitation involves ice at some stage of the process.

Warm Precipitation

Nearly all precipitation begins with condensation of water vapor about small (diameters between

0.00001 to 0.0001 cm) particles in the air called cloud condensation nuclei. Condensation may occur at relative humidities less than 100 percent for hygroscopic particles (those having an affinity for water) or may be delayed until the relative humidity exceeds 100 percent if the particles are hydrophobic (lacking an affinity for water). Sea-salt particles left behind when sea spray evaporates are particularly effective nuclei.

Saturation of air occurs when rising air currents cool adiabatically (that is, without loss of heat) by expansion. Because the saturation vapor pressure of water decreases exponentially with decreasing temperature (a property often summarized by the statement "cold air can hold less water vapor than warm air"), cooling of a moist air mass by lifting is an efficient mechanism for producing saturation and condensation.

The condensation processes are efficient in producing only cloud drops that are too small (diameters between 0.0001 and 0.005 cm) to have an appreciable fall velocity relative to the air (terminal velocity). In order to produce precipitation particles that are heavy enough to fall to the surface, a cloud drop with a radius of 0.001 cm must increase its radius by a factor of 10 and its volume by a factor of 1,000. In clouds with temperatures above freezing, the growth occurs by coalescence, which is simply the merging of water drops that collide. This merging is facilitated when an electric field is present. Laboratory experiments show that drops will bounce off one another in the absence of an electric field.

Cold Precipitation

Whereas collision and coalescence are efficient means for producing precipitation in the warm, humid tropical regions, the formation of precipitation in middle latitudes usually involves ice. Because the vapor pressure at saturation is less over ice than over water, ice crystals will grow at the expense of water drops when both exist together in a supercooled cloud (which contains liquid drops at temperatures below freezing). This mechanism, named the Bergeron-Findeisen process for the scientists who studied it in the 1930s, is a very efficient mechanism for growing ice crystals to a size big enough to fall to the surface.

Although most precipitation in middle latitudes begins as snow at altitudes above the freezing level (about 3 km/1.8 mi.), the form of the precipitation reaching the surface depends on the temperature structure of the atmospheric layers through which the precipitation falls. If the temperature near the ground is warm enough, the snow has time to melt and reaches the ground as rain. A warm layer aloft and a subfreezing layer at the surface may produce sleet (ice pellets) or freezing rain (rain that freezes immediately upon contact with surface objects). Hail occurs when alternating strong updrafts and downdrafts cause ice crystals to pass repeatedly through layers that contain supercooled water. The frequent passage through these layers allows the water to freeze around the growing hailstone and to accumulate in one layer after another. Strong updrafts with velocities of 10 to 30 m/sec (22 to 67 mph) are necessary to prevent the growing hailstone from falling to the surface.

Occurrence

Precipitation is produced whenever moist air rises sufficiently to produce saturation, condensation, and the growth of the precipitation particles. Thunderstorms, with updraft wind components of more than 30 m/sec (67 mph), carry moist low-level air to the tropopause, the upper boundary of the troposphere, and hence can produce heavy precipitation in short amounts of time. Rainfall rates of 10 cm/hr (4 in/hr) are not uncommon in thunderstorms.

Less intense rainfall occurs over much larger scales (horizontal distances of 1,000 km/620 mi. or more) when warm, moist air is lifted over a frontal surface or put in circulation around extratropical cyclones (those poleward of the tropical easterlies). In the Northern Hemisphere the favored location for this frontal type of precipitation is north of the surface warm front and north of the track of an eastward-moving cyclone. Orographic precipitation is produced when air flowing over mountainous terrain is forced to rise. An example is the heavy precipitation that falls on the Sierra Nevadas in western North America. Moist Pacific air forced over the mountains produces up to 250 cm (100 in.) of liquid precipitation annually. As the air descends over the lee slopes and warms adiabatically, the rainfall decreases abruptly. This rain shadow effect can produce extreme variability in rainfall over short horizontal distances.

Geographical Distribution

Because the annual distribution of precipitation depends on small-scale factors such as orographic variability as well as the vertical branches of the average global circulations (the general circulation), the geographic distribution of precipitation is not simple. In general, the average rising branch of the general circulation that occurs in a band between latitudes 15° S and 15° N results in heavy equatorial rainfall. Sinking motion occurs on the average over a band approximately 15° of latitude wide centered at 30° north and south, and rainfall is at a minimum near these latitudes. A second rainfall maximum occurs between 40° and 50° latitude because of the high frequency of traveling cyclones there. Finally, the annual precipitation drops off sharply toward both poles because the cold air cannot contain enough moisture for heavy precipitation.

Hail

Hail is a type of precipitation, or water in the atmosphere. Hail is formed when drops of water freeze together in the cold upper regions of thunderstorm clouds. These chunks of ice are called hailstones. Most hailstones measure between 5 millimeters and 15 centimeters in diameter, and can be round or jagged.

Hailstones are not frozen raindrops. Frozen rain falls as water and freezes as it nears the ground. Hail actually falls as a solid.

Hailstones are formed by layers of water attaching and freezing in a large cloud. A frozen droplet begins to fall from a cloud during a storm, but is pushed back up into the cloud by a strong updraft of wind. When the hailstone is lifted, it hits liquid water droplets. Those droplets then freeze to the hailstone, adding another layer to it. The hailstone eventually falls to Earth when it becomes too heavy to remain in the cloud, or when the updraft stops or slows down.

Certain parts of the world receive more hail than others. The approach of the summer monsoon season in India brings severe thunderstorms, often with tornadoes and hail. A particularly deadly hail storm in Moradabad, India, in 1888 killed more than 250 people. China also experiences frequent hail storms, as do parts of the Midwestern United States. In fact, the Great Plains region of the United States and Canada is called "Hail Alley."

Hail can be transparent (clear) or translucent (cloudy).

Hailstones can cause extreme damage to buildings, vehicles, and crops. Not surprisingly, people have tried to find ways to prevent hail. In the 18th century, Europeans began trying to prevent hail by firing cannons into clouds and ringing church bells.

In the 20th century, Russia and the United States tried cloud seeding. Cloud seeding is adding chemical particles into clouds from rockets or aircraft. Cloud seeding is thought to control rain and hail. There is no clear evidence that any of these techniques are effective.

References

- Heat-wave-causes-list-effect-on-the-economy: thebalance.com, Retrieved 20 August, 2019
- Types of Drought, Drought In-depth: drought.unl.edu, Retrieved 10 February, 2019
- David M. Roth (2002-02-15). "A Fifty year History of Subtropical Cyclones" (PDF). Hydrometeorological Prediction Center. Retrieved 2006-10-04
- Drought-Implications: newworldencyclopedia.org, Retrieved 4 June, 2019
- Tropical-cyclone, science: britannica.com, Retrieved 11 April, 2019
- S. Hastenrath (1991). Climate Dynamics of the Tropics. Springer, pp 244. ISBN 978-0-7923-1346-5. Retrieved on 2009-02-29
- Tornado, science: britannica.com, Retrieved 23 July, 2019
- Acid-rain: ncsu.edu, Retrieved 27 May, 2019
- Thunderstorm, science: britannica.com, Retrieved 7 January, 2019
- How-thundersnow-works: thoughtco.com, Retrieved 4 June, 2019
- Precipitation-weather: scholastic.com, Retrieved 18 March, 2019

3

Atmospheric Science

The branch of science which focuses on the study of the Earth's atmosphere along with the processes related to it is known as atmospheric science. The various areas of interest within this field include atmospheric layers and atmospheric composition. The topics elaborated in this chapter will help in gaining a better perspective about these areas of study within atmospheric science.

Atmospheric science is the term for the study of the atmosphere, its processes, the effects other systems have on the atmosphere, and the effects of the atmosphere on these other systems. Meteorology includes atmospheric chemistry and atmospheric physics with a major focus on weather forecasting. Climatology is the study of atmospheric changes (both long and short-term) that define average climates and their change over time, due to both natural and anthropogenic climate variability. Aeronomy is the study of the upper layers of the atmosphere, where dissociation and ionization are important. Atmospheric science has been extended to the field of planetary science and the study of the atmospheres of the planets of the solar system. Experimental instruments used in atmospheric sciences include satellites, rocketsondes, radiosondes, weather balloons, and lasers.

The atmospheric sciences are traditionally divided into three topical areas—meteorology (the study and forecasting of weather), climatology (the study of long-term atmospheric patterns and their influences), and aeronomy (the study of the physics and chemistry of the upper atmosphere). In meteorology, the focus of study concerns day-to-day and hour-to-hour changes in weather within the lower stratosphere and troposphere.

Within these three major topical areas, the broad nature of the atmospheric sciences has spawned practitioners who specialize in several distinct subfields. Scientists who investigate the physics associated with atmospheric flow are called dynamic meteorologists or simply dynamicists. When the investigation procedure involves the application of large computer models of atmospheric structure and dynamics, the scientists are called numerical modelers.

Atmospheric Layers

The atmosphere is layered, corresponding with how the atmosphere's temperature changes with altitude. By understanding the way temperature changes with altitude, we can learn a lot about how the atmosphere works. While weather takes place in the lower atmosphere, interesting things, such as the beautiful aurora, happen higher in the atmosphere. Why does warm air rise? Gas molecules are able to move freely and if they are uncontained, as they are in the atmosphere, they can take up more or less space.

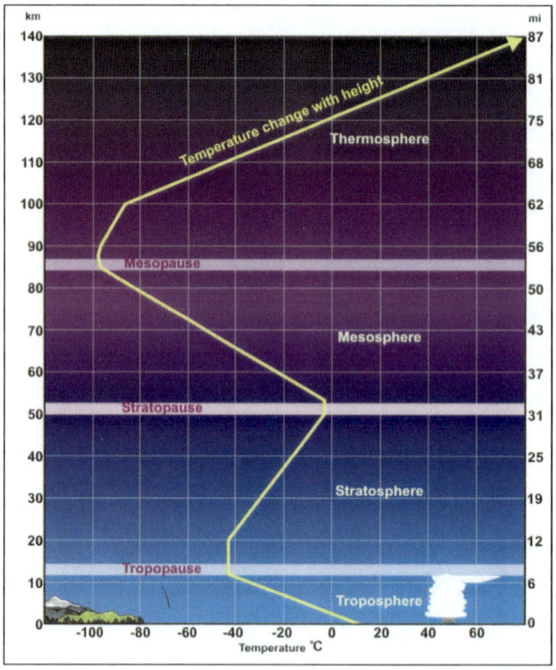

- When gas molecules are cool, they are sluggish and do not take up as much space. With the same number of molecules in less space, both air density and air pressure are higher.
- When gas molecules are warm, they move vigorously and take up more space. Air density and air pressure are lower.

Warmer, lighter air is more buoyant than the cooler air above it, so it rises. The cooler air then sinks down, because it is denser than the air beneath it.

The property that changes most strikingly with altitude is air temperature. Unlike the change in pressure and density, which decrease with altitude, changes in air temperature are not regular. A change in temperature with distance is called a temperature gradient.

The atmosphere is divided into layers based on how the temperature in that layer changes with altitude, the layer's temperature gradient. The temperature gradient of each layer is different. In some layers, temperature increases with altitude and in others it decreases. The temperature gradient in each layer is determined by the heat source of the layer. Most of the important processes of the atmosphere take place in the lowest two layers: the troposphere and the stratosphere.

Troposphere

The temperature of the troposphere is highest near the surface of the Earth and decreases with altitude. On average, the temperature gradient of the troposphere is 6.50 °C per 1,000 m (3.60 °F per 1,000 ft.) of altitude. What is the source of heat for the troposphere?

Earth's surface is a major source of heat for the troposphere, although nearly all of that heat comes from the Sun. Rock, soil, and water on Earth absorb the Sun's light and radiate it back into the atmosphere as heat. The temperature is also higher near the surface because of the greater density of gases. The higher gravity causes the temperature to rise.

Notice that in the troposphere warmer air is beneath cooler air. What do you think the consequence of this is? This condition is unstable. The warm air near the surface rises and cool air higher in the troposphere sinks. So air in the troposphere does a lot of mixing. This mixing causes the temperature gradient to vary with time and place. The rising and sinking of air in the troposphere means that all of the planet's weather takes place in the troposphere.

Sometimes there is a temperature inversion, air temperature in the troposphere increases with altitude and warm air sits over cold air. Inversions are very stable and may last for several days or even weeks. They form:

- Over land at night or in winter when the ground is cold. The cold ground cools the air that sits above it, making this low layer of air denser than the air above it.
- Near the coast where cold seawater cools the air above it. When that denser air moves inland, it slides beneath the warmer air over the land.

Since temperature inversions are stable, they often trap pollutants and produce unhealthy air conditions in cities. At the top of the troposphere is a thin layer in which the temperature does not change with height. This means that the cooler, denser air of the troposphere is trapped beneath the warmer, less dense air of the stratosphere. Air from the troposphere and stratosphere rarely mix.

Stratosphere

Ash and gas from a large volcanic eruption may burst into the stratosphere, the layer above the troposphere. Once in the stratosphere, it remains suspended there for many years because there is so little mixing between the two layers. Pilots like to fly in the lower portions of the stratosphere because there is little air turbulence.

In the stratosphere, temperature increases with altitude. What is the heat source for the stratosphere? The direct heat source for the stratosphere is the Sun. Air in the stratosphere is stable because warmer, less dense air sits over cooler, denser air. As a result, there is little mixing of air within the layer.

The ozone layer is found within the stratosphere between 15 to 30 km (9 to 19 miles) altitudes. The thickness of the ozone layer varies by the season and also by latitude. The ozone layer is extremely important because ozone gas in the stratosphere absorbs most of the Sun's harmful ultraviolet (UV) radiation. Because of this, the ozone layer protects life on Earth. High-energy UV light penetrates cells and damages DNA, leading to cell death (which we know as bad sunburn). Organisms on Earth are not adapted to heavy UV exposure, which kills or damages them. Without the ozone layer to reflect UVC and UVB radiation, most complex life on Earth would not survive long.

Mesosphere

Temperatures in the mesosphere decrease with altitude. Because there are few gas molecules in the mesosphere to absorb the Sun's radiation, the heat source is the stratosphere below. The mesosphere is extremely cold, especially at its top, about −90 degrees C (−130 degrees F).

The air in the mesosphere has extremely low density: 99.9 percent of the mass of the atmosphere is below the mesosphere. As a result, air pressure is very low. A person traveling through the mesosphere would experience severe burns from ultraviolet light since the ozone layer which provides UV protection is in the stratosphere below. There would be almost no oxygen for breathing. Stranger yet, an unprotected traveler's blood would boil at normal body temperature because the pressure is so low.

Thermosphere

The density of molecules is so low in the thermosphere that one gas molecule can go about 1 km before it collides with another molecule. Since so little energy is transferred, the air feels very cold. Within the thermosphere is the ionosphere. The ionosphere gets its name from the solar radiation that ionizes gas molecules to create a positively charged ion and one or more negatively charged electrons. The freed electrons travel within the ionosphere as electric currents. Because of the free ions, the ionosphere has many interesting characteristics. At night, radio waves bounce off the ionosphere and back to Earth. This is why you can often pick up an AM radio station far from its source at night.

The Van Allen radiation belts are two doughnut-shaped zones of highly charged particles that are located beyond the atmosphere in the magnetosphere. The particles originate in solar flares and fly to Earth on the solar wind. Once trapped by Earth's magnetic field, they follow along the field's magnetic lines of force. These lines extend from above the equator to the North Pole and also to the South Pole then return to the equator.

When massive solar storms cause the Van Allen belts to become overloaded with particles, the result is the most spectacular feature of the ionosphere—the night time aurora. The particles spiral along magnetic field lines toward the poles. The charged particles energize oxygen and nitrogen gas molecules, causing them to light up. Each gas emits a particular color of light.

There is no real outer limit to the exosphere, the outermost layer of the atmosphere; the gas molecules finally become so scarce that at some point there are no more. Beyond the atmosphere is the solar wind. The solar wind is made of high-speed particles, mostly protons and electrons, traveling rapidly outward from the Sun.

Atmospheric Physics

Atmospheric physics is the application of physics to the study of the atmosphere. Atmospheric physicists attempt to model Earth's atmosphere and the atmospheres of the other planets using fluid flow equations, chemical models, radiation balancing, and energy transfer processes in the atmosphere and underlying oceans. In order to model weather systems, atmospheric physicists employ elements of scattering theory, wave propagation models, cloud physics, statistical mechanics and spatial statistics which are highly mathematical and related to physics. It has close links to meteorology and climatology and also covers the design and construction of instruments for studying the atmosphere and the interpretation of the data they provide, including remote sensing instruments.

Atmospheric Temperature

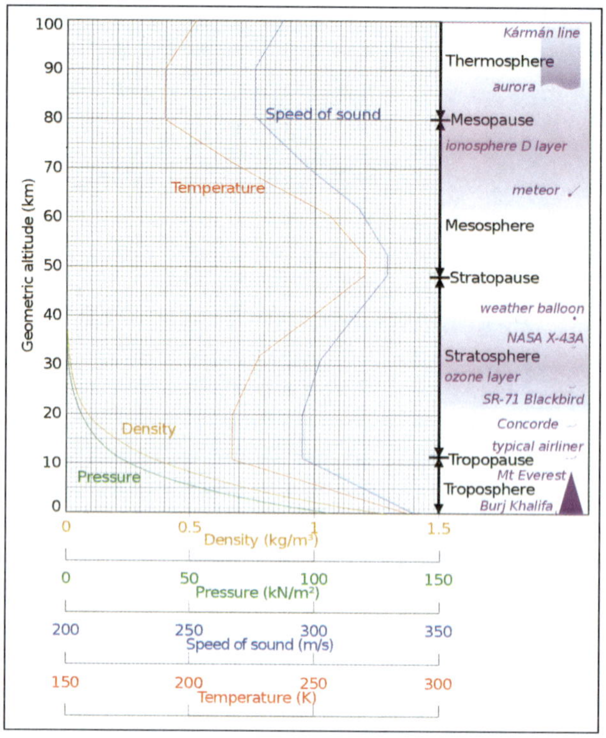

Atmospheric temperature is a function of the modification of solar radiant energy by air, clouds, land, sea and other water surfaces.

The temperature of Earth's atmosphere varies with the distance from the equator (latitude) and height above the surface (altitude). It also changes with time, varying from season to season, and from day to night, as well as irregularly due to passing weather systems. If local variations are averaged out on a global basis, however, a pattern of global average temperatures emerges.

Global Variation of Mean Temperature

Global variations of average surface-air temperatures are largely due to latitude, continentality, ocean currents, and prevailing winds.

The effect of latitude is evident in the large north-south gradients in average temperature that occur at middle and high latitudes in each winter hemisphere. These gradients are due mainly to the rapid decrease of available solar radiation but also in part to the higher surface reflectivity at high latitudes associated with snow and ice and low solar elevations. A broad area of the tropical ocean, by contrast, shows little temperature variation.

Continentality is a measure of the difference between continental and marine climates and is mainly the result of the increased range of temperatures that occurs over land compared with water. This difference is a consequence of the much lower effective heat capacities of land surfaces as well as of their generally reduced evaporation rates. Heating or cooling of a land surface takes place in a thin layer, the depth of which is determined by the ability of the ground to conduct heat. The greatest temperature changes occur for dry, sandy soils, because they are poor conductors with very small effective heat capacities and contain no moisture for evaporation. By far the greatest effective heat capacities are those of water surfaces, owing to both the mixing of water near the surface and the penetration of solar radiation that distributes heating to depths of several metres. In addition, about 90 percent of the radiation budget of the ocean is used for evaporation. Ocean temperatures are thus slow to change.

The effect of continentality may be moderated by proximity to the ocean, depending on the direction and strength of the prevailing winds. Contrast with ocean temperatures at the edges of each continent may be further modified by the presence of a north- or south-flowing ocean current. For most latitudes, however, continentality explains much of the variation in average temperature at fixed latitude as well as variations in the difference between January and July temperatures.

Diurnal, Seasonal and Extreme Temperatures

The diurnal range of temperature generally increases with distance from the sea and toward those places where solar radiation is strongest—in dry tropical climates and on high mountain plateaus (owing to the reduced thickness of the atmosphere to be traversed by the Sun's rays). The average difference between the day's highest and lowest temperatures is 3 °C (5 °F) in January and 5 °C (9 °F) in July in those parts of the British Isles nearest the Atlantic. The difference is 4.5 °C (8 °F) in January and 6.5 °C (12 °F) in July on the small island of Malta. At Tashkent, Uzbekistan, it is 9 °C (16 °F) in January and 15.5 °C (28 °F) in July, and at Khartoum, Sudan, the corresponding figures are 17 °C (31 °F) and 13.5 °C (24 °F). At Kandahār, Afghanistan, which lies more than 1,000 metres (about 3,300 feet) above sea level, it is 14 °C (25 °F) in January and 20 °C (36 °F) in July. There, the average difference between the day's highest and lowest temperatures exceeds 23 °C (41 °F) in September and October, when there is less cloudiness than in July. Near the ocean at Colombo, Sri L., the figures are 8 °C (14 °F) in January and 4.5 °C (8 °F) in July.

The seasonal variation of temperature and the magnitudes of the differences between the same month in different years and different epochs generally increase toward high latitudes and with distance from the ocean. Extreme temperatures observed in different parts of the world are listed in the table.

World temperature extremes			
Highest recorded air temperature			
		Temperature	
Continent or Region	Place (with elevation)	Degrees C	Degrees F
Africa	Kebili, Tunisia (38.1 m or 125 ft)	55	131
Antarctica	Vanda Station 77°32′ S 161°40′ E (15 m or 49 ft)	15	59
Asia	Tirat Zevi, Israel (−220 m or −722 ft)	54	129.2
Australia	Oodnadatta, South Australia (112 m or 367 ft)	50.7	123
Europe	Athens, Greece (236 m or 774 ft)	48	118.4
North America	Death Valley (Greenland Ranch), California, U.S. (−54 m or −177 ft)	56.7	134

South America	Rivadavia, Argentina (668 m or 2,192 ft)	48.9	120
Oceania	Tuguegarao, Luzon, Philippines (62 m or 203 ft)	42.2	108
Lowest recorded air temperature			
		Temperature	
Continent or Region	Place (with elevation*)	Degrees C	Degrees F
Africa	Ifrane, Morocco (1,635 m or 5,364 ft)	−23.9	−11
Antarctica	Vostok 77°32′ S 106°40′ E (3,420 m or 11,220 ft)	−89.2	−128.6
Asia	Verkhoyansk, Russia (107 m or 351 ft) Oymyakon, Russia (800 m or 2,624 ft)	−67.8	−90
Australia	Charlotte Pass, New South Wales (1,755 m or 5,758 ft)	−23	−9.4
Europe	Ust-Shchuger, Russia (85 m or 279 ft)	−58.1	−72.6
North America	Snag, Yukon, Canada (646 m or 2,119 ft)	−63	−81.4
South America	Sarmiento, Argentina (268 m or 879 ft)	−32.8	−27
Above or below sea level.			

Variation with Height

There are two main levels where the atmosphere is heated—namely, at Earth's surface and at the top of the ozone layer (about 50 km, or 30 miles, up) in the stratosphere. Radiation balance shows a net gain at these levels in most cases. Prevailing temperatures tend to decrease with distance from these heating surfaces (apart from the ionosphere and the outer atmospheric layers, where other processes are at work). The world's average lapse rate of temperature (change with altitude) in the lower atmosphere is 0.6 to 0.7 °C per 100 metres (about 1.1 to 1.3 °F per 300 feet). Lower temperatures prevail with increasing height above sea level for two reasons: because there is a less favourable radiation balance in the free air, and because rising air—whether lifted by convection currents above a relatively warm surface or forced up over mountains—undergoes a reduction of temperature associated with its expansion as the pressure of the overlying atmosphere declines. This is the adiabatic lapse rate of temperature, which equals about 1 °C per 100 metres (about 2 °F per 300 feet) for dry air and 0.5 °C per 100 metres (about 1 °F per 300 feet) for saturated air, in which condensation (with liberation of latent heat) is produced by adiabatic cooling. The difference between these rates of change of temperature (and therefore density) of rising air currents and the state of the surrounding air determines whether the upward currents are accelerated or retarded—i.e., whether the air is unstable, so vertical convection with its characteristically attendant tall cumulus cloud and shower development is encouraged or whether it is stable and convection is damped down.

For these reasons, the air temperatures observed on hills and mountains are generally lower than on low ground, except in the case of extensive plateaus, which present a raised heating surface (and on still, sunny days, when even a mountain peak is able to warm appreciably the air that remains in contact with it).

Circulation, Currents and Ocean-atmosphere Interaction

The circulation of the ocean is a key factor in air temperature distribution. Ocean currents that have a northward or southward component, such as the warm Gulf Stream in the North Atlantic or the cold Peru (Humboldt) Current off South America, effectively exchange heat between low and high latitudes. In tropical latitudes the ocean accounts for a third or more of the pole ward heat transport; at latitude 50 °N, the ocean's share is about one-seventh. In the particular sectors where the currents are located, their importance is of course much greater than these figures, which represent hemispheric averages.

A good example of the effect of a warm current is that of the Gulf Stream in January, which causes a strong east-west gradient in temperatures across the eastern edge of the North American continent. The relative warmth of the Gulf Stream affects air temperatures all the way across the Atlantic, and prevailing westerlies extend the warming effect deep into northern Europe. As a result, January temperatures of Tromsø, Nor. (69°40′ N), for example, average 24 °C (43 °F) above the mean for that latitude. The Gulf Stream maintains a warming influence in July, but it is not as noticeable because of the effects of continentality.

The ocean, particularly in areas where the surface is warm, also supplies moisture to the atmosphere. This in turn contributes to the heat budget of those areas in which the water vapour is condensed into clouds, liberating latent heat in the process. This set of events occurs frequently in high latitudes and in locations remote from the ocean where the moisture was initially taken up.

The great ocean currents are themselves wind-driven—set in motion by the drag of the winds over vast areas of the sea surface, especially where the tops of waves increase the friction with the air above. At the limits of the warm currents, particularly where they abut directly upon a cold current—as at the left flank of the Gulf Stream in the neighbourhood of the Grand Banks off Newfoundland and at the subtropical and Antarctic convergences in the oceans of the Southern Hemisphere—the strong thermal gradients in the sea surface result in marked differences in the heating of the atmosphere on either side of the boundary. These temperature gradients tend to position and guide the strongest flow of the jet stream in the atmosphere above and thereby influence the development and steering of weather systems.

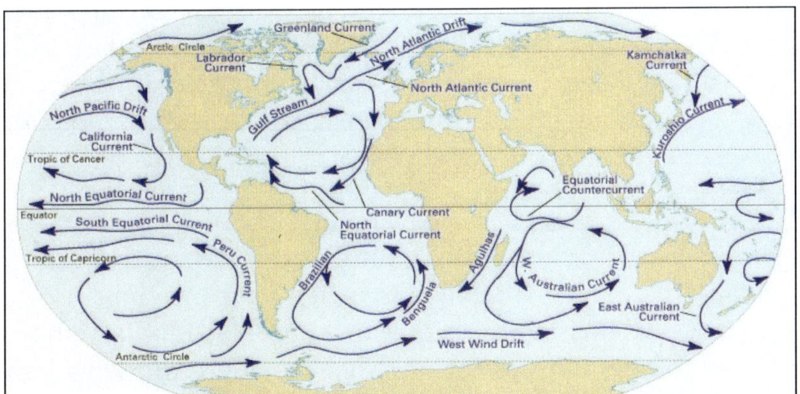

Major surface currents of the world's oceans. Subsurface currents also move vast amounts of water, but they are not known in such detail.

Interactions between the ocean and the atmosphere proceed in both directions. They also operate at different rates. Some interesting lag effects, which are of value in long-range weather forecasting, arise through the considerably slower circulation of the ocean. Thus, enhanced strength of the easterly trade winds over low latitudes of the Atlantic north and south of the Equator impels more water toward the Caribbean and the Gulf of Mexico, producing a stronger flow and greater warmth in the Gulf Stream approximately six months later. Anomalies in the position of the Gulf Stream–Labrador

Current boundary, which produce a greater or lesser extent of warm water near the Grand Banks, so affect the energy supply to the atmosphere and the development and steering of weather systems from that region that they are associated with rather persistent anomalies of weather pattern over the British Isles and northern Europe. Anomalies in the equatorial Pacific and in the northern limit of the Kuroshio Current (also called the Japan Current) seem to have effects on a similar scale. Indeed, through their influence on the latitude of the jet stream and the wavelength (that is, the spacing of cold trough and warm ridge regions) in the upper westerlies, these ocean anomalies exercise an influence over the atmospheric circulation that spreads to all parts of the hemisphere.

Sea-surface temperature anomalies that recur in the equatorial Pacific at variable intervals of two to seven years can sometimes produce major climatic perturbations. One such anomaly is known as El Niño (Spanish for "The Child"; it was so named by Peruvian fishermen who noticed its onset during the Christmas season).

During an El Niño event, warm surface water flows eastward from the equatorial Pacific, in at least partial response to weakening of the equatorial easterly winds, and replaces the normally cold upwelling surface water off the coast of Peru and Ecuador that is associated with the northward propagation of the cold Peru Current. The change in sea-surface temperature transforms the coastal climate from arid to wet. The event also affects atmospheric circulation in both hemispheres and is associated with changes in precipitation in regions of North America, Africa, and the western Pacific.

Short-term Temperature Changes

Many interesting short-term temperature fluctuations also occur, usually in connection with local weather disturbances. The rapid passage of a mid-latitude cold front, for example, can drop temperatures by 10 °C (18 °F) in a few minutes and, if followed by the sustained movement of a cold air mass, by as much as 50 °C in 24 hours, with life-threatening implications for the unwary. Temperature increases of up to 40 °C in a few hours also are possible downwind of major mountain ranges when air that has been warmed by the release of latent heat on the windward side of a range is forced to descend rapidly on the other side (such a wind is variously called chinook, foehn, or Santa Ana). Changes of this kind, however, involve a wider range of meteorological processes.

Atmospheric Temperature Range

Atmospheric temperature range is the numerical difference between the minimum and maximum values of temperature observed in a given location. A temperature range may refer to a period of time or to an average. The variation in temperature that occurs from the highs of the day to the cool of nights is called diurnal temperature variation. The size of ground-level atmospheric temperature ranges depends on several factors, such as:

- The average temperature,
- The average humidity,
- The regime of winds,
- The proximity to large bodies of water, such as the sea.

A location which combines an average temperature of 19 degrees Celsius, 60% average humidity and a temperature range of about 10 degrees Celsius around the average temperature is considered ideal in terms of comfort for the human species. Most of the places with these characteristics are located in the transition between temperate and tropical climates, approximately around the tropics, particularly in the Southern hemisphere.

Upper Air Temperature

The term "upper air" refers to the part of the atmosphere well above Earth's surface. The temperature of the atmosphere in this region is a fundamental climate variable. Recent changes in atmospheric temperature have been attributed to human-induced climate change. Continued monitoring of atmospheric temperature is critical to advancing our understanding of the sensitivity of Earth's climate to changing atmospheric composition.

Measurement Methods for Upper Air Temperatures

There are several methods available for the measurement of the upper air.

- Radiosondes (commonly called weather balloons): These are small instruments lifted aloft by helium-filled balloons. Measurements made by temperature, pressure, and humidity sensors are radioed back to the surface.
    - Advantages: Direct measurement of temperature, high vertical resolution.
    - Disadvantages: Limited spatial sampling, calibration problems with changing instruments and methods.
- Microwave Sounders: These are satellite-borne instruments that measure the radiance of Earth at microwave frequencies, which allows scientists to deduce the temperature of thick atmospheric layers.
    - Advantages: Global coverage at a high sampling rate.
    - Disadvantages: Coarse vertical resolution.
- Infrared Sounders: These are satellite-borne instruments that measure the radiance of Earth at infrared frequencies, which allows scientists to retrieve the temperature of thick atmospheric layers using inversion algorithms.
    - Advantages: Global coverage at a high sampling rate, moderate vertical resolution.
    - Disadvantages: Sensitivity to contamination by clouds and other aerosols.
- GPS Radio Occultation: The approach uses satellite-borne GPS receivers to measure the refraction of the GPS signals by the Earth's atmosphere. This allows for the retrieval of vertical temperature and moisture profiles.
    - Advantages: Absolute calibration, high vertical resolution, global coverage.
    - Disadvantages: Only about 10 years of data, less frequent sampling compared to satellite-borne sounders.

Atmospheric Pressure

The atmospheric pressure is the weight exerted by the overhead atmosphere on a unit area of surface. It can be measured with a mercury barometer, consisting of a long glass tube full of mercury inverted over a pool of mercury:

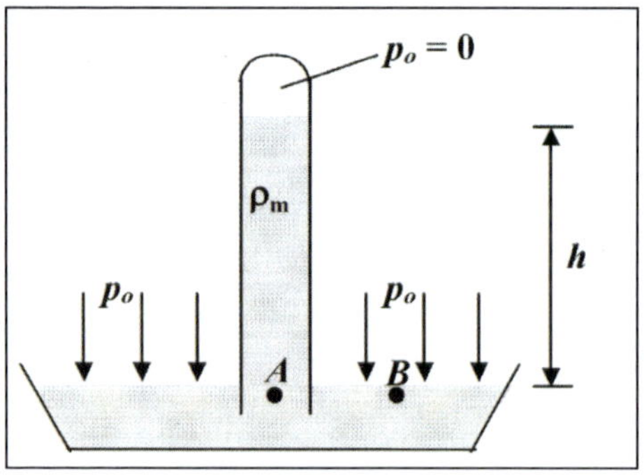

Mercury barometer

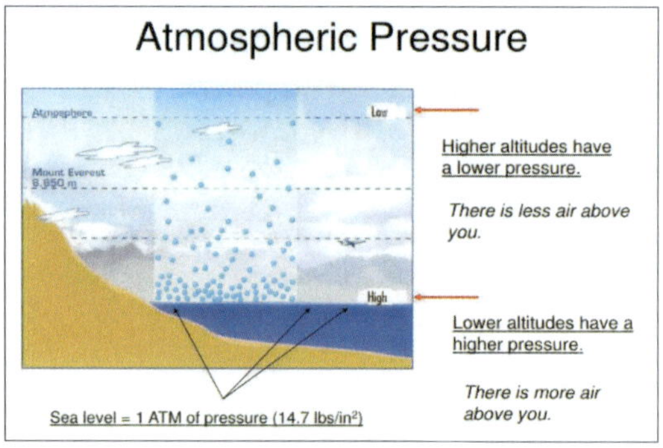

When the tube is inverted over the pool, mercury flows out of the tube, creating a vacuum in the head space, and stabilizes at an equilibrium height h over the surface of the pool. This equilibrium requires that the pressure exerted on the mercury at two points on the horizontal surface of the pool, A (inside the tube) and B (outside the tube), be equal. The pressure P_A at point A is that of the mercury column overhead, while the pressure P_B at point B is that of the atmosphere overhead. We obtain P_A from measurement of h:

$$P_A = \rho_{Hg} gh$$

where, ρ_{Hg} = 13.6 g cm^{-3} is the density of mercury and g = 9.8 m s^{-2} is the acceleration of gravity. The mean value of h measured at sea level is 76.0 cm, and the corresponding atmospheric pressure is 1.013 x 10^5 kg m^{-1} s^{-2} in SI units. The SI pressure unit is called the Pascal (Pa); 1 Pa = 1 kg m^{-1} s^{-2}. Customary pressure units are the atmosphere (atm) (1 atm = 1.013 x 10^5 Pa), the bar (b) (1 b = 1 x 10^5 Pa), the millibar (mb) (1 mb = 100 Pa), and the torr (1 torr = 1 mm Hg = 134 Pa). The use of millibars is slowly giving way to the equivalent SI unit of hectoPascals (hPa). The mean atmospheric pressure at sea level is given equivalently as P = 1.013 x 10^5 Pa = 1013 hPa = 1013 mb = 1 atm = 760 torr.

Mass of the Atmosphere

The global mean pressure at the surface of the Earth is P_S = 984 hPa, slightly less than the mean sea-level pressure because of the elevation of land. We deduce the total mass of the atmosphere m_a:

$$m_a = \frac{4\pi R^2 P_S}{g} = 5.2x10^{18}\ kg$$

where, R = 6400 km is the radius of the Earth. The total number of moles of air in the atmosphere is N_a = ma/M_a = 1.8 x 10^{20} moles.

Vertical Profiles of Pressure and Temperature

Figure shows typical vertical profiles of pressure and temperature observed in the atmosphere.

Pressure decreases exponentially with altitude. The fraction of total atmospheric weight located above altitude z is P(z)/P(0). At 80 km altitude the atmospheric pressure is down to 0.01 hPa, meaning that 99.999% of the atmosphere is below that altitude. You see that the atmosphere is of relatively thin vertical extent. Astronomer Fred Hoyle once said, "Outer space is not far at all; it's only one hour away by car if your car could go straight up."

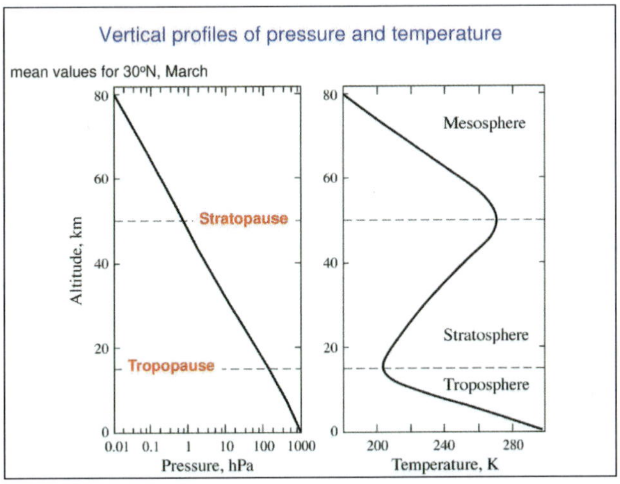

Mean pressure and temperature vs. altitude at 30 °N, March.

Atmospheric scientists partition the atmosphere vertically into domains separated by reversals of the temperature gradient, as shown in figure. The troposphere extends from the surface to 8-18 km altitude depending on latitude and season. It is characterized by a decrease of temperature with altitude which can be explained simply though not quite correctly by solar heating of the surface. The stratosphere extends from the top of the troposphere (the tropopause) to about 50 km altitude (the stratopause) and is characterized by an increase of temperature with altitude due to absorption of solar radiation by the ozone layer. In the mesosphere, above the ozone layer, the temperature decreases again with altitude. The mesosphere extends up to 80 km (mesopause) above which lays the thermosphere where temperatures increase again with altitude due to absorption of strong UV solar radiation by N_2 and O_2. The troposphere and stratosphere account together for 99.9% of total atmospheric mass and are the domains of main interest from an environmental perspective.

Barometric Law

We focus here on explaining the vertical profile of pressure. Consider an elementary slab of atmosphere (thickness dz, horizontal area A) at altitude z:

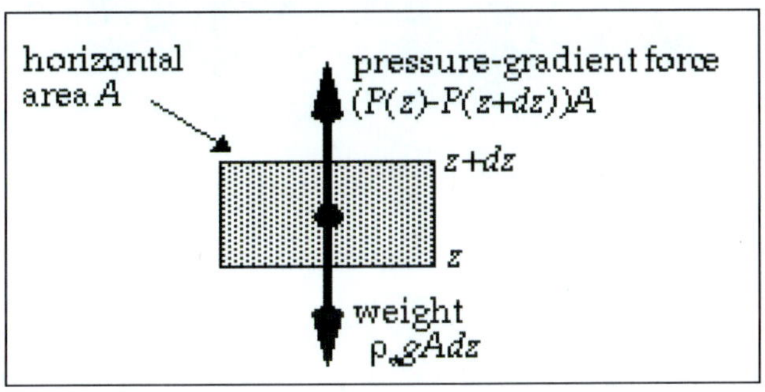

Vertical forces acting on an elementary slab of atmosphere.

The atmosphere exerts an upward pressure force P(z)A on the bottom of the slab and a downward pressure force P(z + dz)A on the top of the slab; the net force, (P(z)-P(z + dz))A, is called the pressure-gradient force. Since P(z) > P(z + dz), the pressure-gradient force is directed upwards. For the slab to be in equilibrium, its weight must balance the pressure-gradient force:

$$\rho_a gAdz = \left(P(z) - P(z+dz)\right)A$$

Rearranging yields,

$$\frac{P(z+dz) - P(z)}{dz} = -\rho_a g$$

The left hand side is dP/dz by definition. Therefore,

$$\frac{dP}{dz} = -\rho_a g$$

Now, from the ideal gas law,

$$\rho_a = \frac{\mathrm{PM_a}}{\mathrm{RT}}$$

where, M_a is the molecular weight of air and T is the temperature. Substituting $\left(\rho_a = \frac{\mathrm{PM_a}}{\mathrm{RT}}\right)$ into $\left(\frac{dP}{dz} = -\rho_a g\right)$ yields:

$$\frac{dP}{P} = -\frac{M_a g}{RT} dz$$

We now make the simplifying assumption that T is constant with altitude; as shown in figure, T varies by only 20% below 80 km. We then integrate $\left(\frac{dP}{P} = -\frac{M_a g}{RT} dz\right)$ to obtain,

$$In\ P(z) - In\, P(0) = -\frac{M_a g}{RT} z$$

which is equivalent to,

$$P(z) = P(0)\exp\left(-\frac{M_a g}{RT} z\right)$$

Equation $P(z) = P(0)\exp\left(-\frac{M_a g}{RT} z\right)$ is called the barometric law. It is convenient to define a scale height H for the atmosphere:

$$H = \frac{RT}{M_a g}$$

leading to a compact form of the Barometric Law:

$$P(z) = P(0)e^{-\frac{z}{H}}$$

For a mean atmospheric temperature T = 250 K the scale height is H = 7.4 km. The barometric law explains the observed exponential dependence of P on z in figure; from equation $\left(P(z) = P(0)e^{-\frac{z}{H}}\right)$, a plot of z vs. ln P yields a straight line with slope -H (check out that the slope in Figure is indeed close to -7.4 km). The small fluctuations in slope in Figure are caused by variations of temperature with altitude which we neglected in our derivation.

The vertical dependence of the air density can be similarly formulated. From

$$\left(\rho_a = \frac{PM_a}{RT}\right),$$

ρ_a and P are linearly related if T is assumed constant, so that,

$$\rho_a(z) = \rho_a(0)e^{-\frac{z}{H}}.$$

A similar equation applies to the air number density n_a. For every H rise in altitude, the pressure and density of air drop by a factor e = 2.7; thus H provides a convenient measure of the thickness of the atmosphere.

In calculating the scale height from $\left(H = \frac{RT}{M_a g}\right)$ we assumed that air behaves as a homogeneous gas of molecular weight M_a = 29 g mol^{-1}. Dalton's law stipulates that each component of the air mixture must behave as if it were alone in the atmosphere. One might then expect different components to have different scale heights determined by their molecular weight. In particular, considering the difference in molecular weight between N_2 and O_2, one might expect the O_2 mixing ratio to decrease with altitude. However, gravitational separation of the air mixture takes place by molecular diffusion, which is considerably slower than turbulent vertical mixing of air for altitudes below 100 km. Turbulent mixing thus maintains a homogeneous lower atmosphere. Only above 100 km does significant gravitational separation of gases begin to take place, with lighter gases being enriched at higher altitudes. During the debate over the harmful effects of chlorofluorocarbons (CFCs) on stratospheric ozone, some not-so-reputable scientists claimed that CFCs could not possibly reach the stratosphere because of their high molecular weights and hence low scale heights. In reality, turbulent mixing of air ensures that CFC mixing ratios in air entering the stratosphere are essentially the same as those in surface air.

Sea-breeze Circulation

An illustration of the Barometric Law is the sea-breeze circulation commonly observed at the beach on summer days. Consider a coastline with initially the same atmospheric temperatures and pressures over land (L) and over sea (S). Assume that there is initially no wind. In summer during the day the land surface is heated to a higher temperature than the sea. This difference is due in part to the larger heat capacity of the sea, and in part to the consumption of heat by evaporation of water.

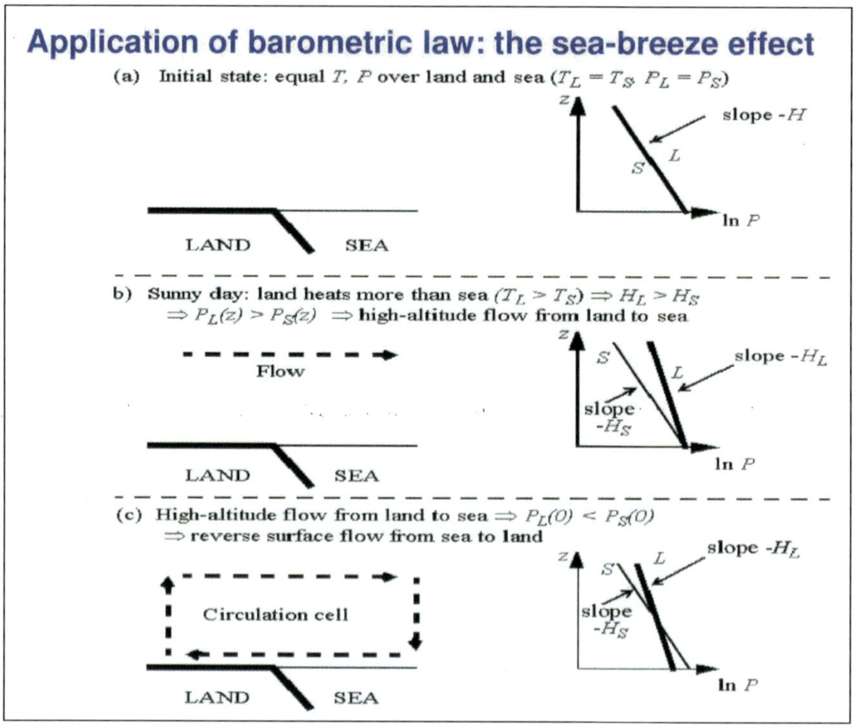

The sea-breeze circulation.

As long as there is no flow of air between land and sea, the total air columns over each region remain the same so that at the surface $P_L(0) = P_S(0)$. However, the higher temperature over land results in a larger atmospheric scale height over land ($H_L > H_S$), so that above the surface $P_L(z) > P_S(z)$. This pressure difference causes the air to flow from land to sea, decreasing the mass of the air column over the land; consequently, at the surface, $P_L(0) < P_S(0)$ and the wind blows from sea to land (the familiar "sea breeze"). Compensating vertical motions result in the circulation cell shown in figure. This cell typically extends ~10 km horizontally across the coastline and ~1 km vertically. At night a reverse circulation is frequently observed (the land breeze) as the land cools faster than the sea.

Atmospheric Chemistry

Atmospheric chemistry is a branch of atmospheric science in which the chemistry of the Earth's atmosphere and that of other planets is studied. It is a multidisciplinary field of research and draws on environmental chemistry, physics, meteorology, computer modeling, oceanography, geology and volcanology and other disciplines. Research is increasingly connected with other areas of study such as climatology. The composition and chemistry of the atmosphere is of importance for several reasons, but primarily because of the interactions between the atmosphere and living organisms. The composition of the Earth's atmosphere has been changed by human activity and some of these changes are harmful to human health, crops and ecosystems. Examples of problems which have been addressed by atmospheric chemistry include acid rain, photochemical smog and global warming. Atmospheric chemistry seeks to understand the causes of these problems, and by obtaining a theoretical understanding of them, allow possible solutions to be tested and the effects of changes in government policy evaluated.

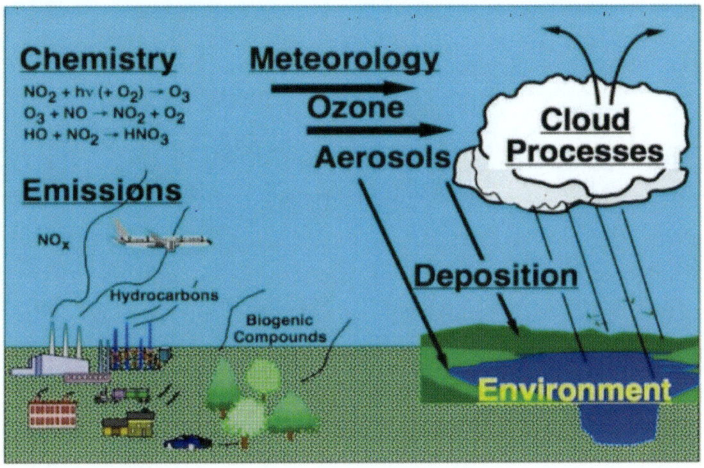

Atmospheric Composition

Earth's atmosphere is composed of about 78% nitrogen, 21% oxygen, and 0.93% argon. The remainder, less than 0.1%, contains many small but important trace gases, including water vapor, carbon dioxide, and ozone. All of these trace gases have important effects on the earth's climate. The atmosphere can be divided into vertical layers determined by the way temperature changes with height. The layer closest to the surface is the troposphere, which contains over 80% of the atmospheric mass and nearly all the water vapor. The next layer, the stratosphere, contains most of the atmosphere's ozone, which absorbs high energy radiation from the sun and makes life on the surface possible. Above the stratosphere are the mesosphere and thermosphere. These two layers include regions of charged atoms and molecules, or ions. Called the ionosphere, this region is important to radio communications, since radio waves can bounce off the layer and travel great distances. It is thought that the present atmosphere developed from gases ejected by volcanoes. Oxygen, upon which all animal life depends, probably built up as excess emissions from plants that produce it as a waste product during photosynthesis. Human activities may be affecting the levels of some important atmospheric components, particularly carbon dioxide and ozone.

Major Gases

The most common atmospheric gas, nitrogen (chemical symbol N_2) accounts for about 78% of the atmosphere. Nitrogen gas is largely inert, meaning that it does not readily react with other substances to form new chemical compounds. The next most common gas, oxygen (O_2), makes up about 21% of the atmosphere. Oxygen is required for the respiration (breathing) of all animal life on Earth, from humans to bacteria. In contrast to nitrogen, oxygen is extremely reactive. It participates in oxidation, a type of chemical reaction that can be observed everywhere. Some common examples of oxidation are apples turning from white to brown after being sliced, the rusting of iron, and the very rapid oxidation reaction we call fire. Just under 1% of the atmosphere is made up of argon (Ar), which is a very inert noble gas, meaning that it does not take part in any chemical reactions under normal circumstances.

Together, these three gases account for 99.96% of the atmosphere. The remaining 0.04% contains a wide variety of trace gases, several of which are crucial to life on Earth.

Important Trace Gases

Carbon dioxide (CO_2) affects the earth's climate and plays a large support role in the biosphere, the collection of living things that populate the earth's surface. Only about 0.0325% of the atmosphere is CO_2. Carbon dioxide is required by plant life for photosynthesis, the process of using sunlight to store energy as simple sugars, upon which all life on Earth depends. Carbon dioxide is also one of a class of compounds called greenhouse gases. These gases are made up of molecules that absorb and emit infrared radiation, which we feel as heat. The solar energy radiated from the sun is mostly in the visible range, within a narrow band of wavelengths. This radiation is absorbed by the earth's surface, and then reradiated back out to space not as visible light, but as longer wavelength infrared radiation. Greenhouse gas molecules absorb some of this radiation before it escapes to space, and re-emit some of it back toward the surface. In this way, these gases trap some of the escaping heat and increase the overall temperature of the atmosphere. If the atmosphere had no greenhouse gases, it is estimated that the earth's surface would be 90 °F (32 °C) cooler.

Water vapor (H_2O) is found in the atmosphere in small and highly variable amounts. While it is nearly absent in most of the atmosphere, its concentration can range up to 4% in very warm, humid areas close to the surface. Despite its relative scarcity, atmospheric water probably has more of an impact on the earth than any of the major gases, aside from oxygen. Water vapor participates in the hydrologic cycle, the process that moves water between the oceans, the land surface waters, the atmosphere, and the polar ice caps. This water cycling drives erosion and rock weathering, determines the earth's weather, and sets up climate conditions that make land areas dry or wet, habitable or inhospitable. When cooled sufficiently, water vapor forms clouds by condensing to liquid water droplets, or at lower temperatures, solid ice crystals. Besides creating rain or snow, clouds affect Earth's climate by reflecting some of the energy coming from the sun, making the planet somewhat cooler. Water vapor is also an important greenhouse gas. It is concentrated near the surface and is much more prevalent near the tropics than in the Polar Regions.

Ozone (O_3) is almost all found in a layer about 9–36 mi (15–60 km) in attitude. Ozone gas is irritating to peo ples' eyes and skin, and chemically attacks rubber and plant tissue. Nevertheless, it is vital to life on Earth because it absorbs most of the high energy radiation from the sun that is harmful to plants and animals. A portion of the energy radiated by the sun lies in the ultraviolet (UV) region. This shorter wavelength radiation is responsible for suntans, and is sufficiently powerful to harm cells, cause skin cancer, and burn tissue, as anyone who has had painful sunburn knows. The ozone molecules, along with molecules of O_2, absorb nearly all the high energy UV rays, protecting the earth's surface from the most damaging radiation. The first step in this process occurs high in the atmosphere, where O_2 molecules absorb very high energy UV radiation. Upon doing so, each absorbing molecule breaks up into two oxygen atoms. The oxygen atoms eventually collide with another O_2molecule, forming a molecule of ozone, O_3 (a third molecule is required in the collision to carry away excess energy). Ozone in turn may absorb UV of slightly longer wavelength, which knocks off one of its oxygen atoms and leaves O_2. The free oxygen atom, being very reactive, will almost immediately recombine with another O_2, forming more ozone. The last two steps of this cycle keep repeating but do not create any new chemical compounds; they only act to absorb ultraviolet radiation. The amount of ozone in the stratosphere is minute. If it were all transported to the surface, the ozone gas would form a layer about 0.1–0.16 in (2.5–4.0 mm) thick. This layer, as thin as it is, is sufficient to shield the earth's occupants from harmful solar radiation.

Aerosols

In addition to gases, the atmosphere has a wide variety of tiny particles suspended in the air, known collectively as aerosols. These particles may be liquid or solid, and are so small that they may require very long times to settle out of the atmosphere by gravity. Examples of aerosols include bits of suspended soil or desert sand, tiny smoke particles from a forest fire, salt particles left over after a droplet of ocean water has evaporated, plant pollen, volcanic dust plumes, and particles formed from the pollution created by a coal burning power plant. Aerosols significantly affect the atmospheric heat balance, cloud growth, and optical properties.

Aerosols cover a very wide size range. Raindrops suspended in a cloud are about 0.04–0.24 in (1–6 mm) in diameter. Fine desert sand and cloud droplets range in diameter down to about 0.0004 in (0.01 mm). Sea salt particles and smoke particles are 1/100th of this, about 0.0001 mm, or 0.1 micrometer, in diameter (1 micrometer = one thousandth of a millimeter). Smallest of all are the particles that form when certain gases condense; that is, when several gas molecules come together to form a stable cluster. These are the Aitkin nuclei, whose diameters can be measured down to a few nanometers (1 nanometer = one millionth of a millimeter).

Some aerosols are just the right size to efficiently scatter sunlight, making the atmosphere look hazy. Under the right conditions, aerosols act as collecting points for water vapor molecules, encouraging the growth of cloud droplets and speeding the formation of clouds. They may also play a role in Earth's climate; the aerosols are known to reflect a portion of incoming solar radiation back to space, which lowers the temperature of the earth's surface. Current research is focused on estimating how much cooling is provided by aerosols, as well as how and when aerosols form in the atmosphere.

Carbon Cycle

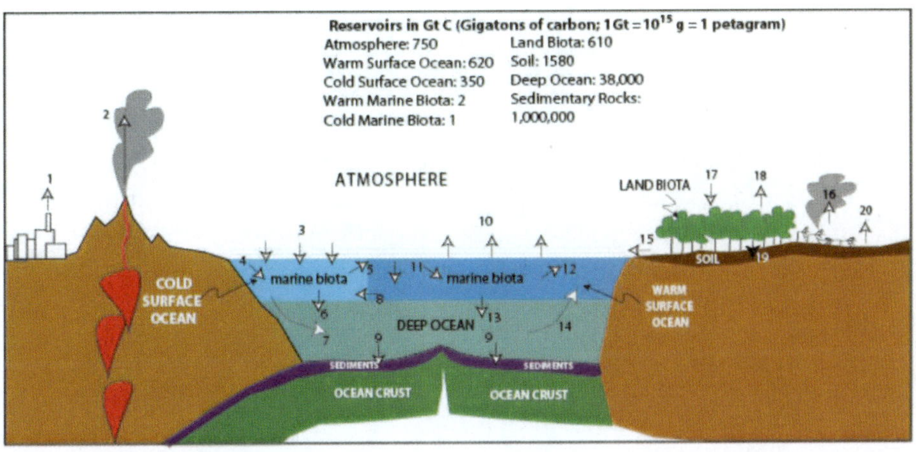

Carbon is a fundamental building block of life; life on Earth is comprised of carbon-based life forms. Carbon also cycles through the oceans and the biosphere over both short and long-term time scales. The geological carbon cycle takes place over hundreds of millions of years and involves the cycling of carbon through the various layers of the Earth. The biological/physical carbon cycle occurs over days, weeks, months, and years and involves the absorption, conversion, and respiration of carbon by living organisms.

Processes of Carbon Flow in the Terrestrial Realm

Photosynthesis

Since its invention over 3 billion years ago, photosynthesis has been one of the most important processes on Earth, helping to make our planet different from all the other planets. The basic idea is that plants capture light energy and use it to split water molecules and then combine the products with carbon dioxide to make carbohydrates, which are used for fuel and construction of the plant; oxygen is a by-product of this reaction, which is summarized as:

$$CO_2 + H_2O + \text{light energy} \rightarrow CH_2O + O_2$$

This process takes places in the chloroplasts located in the interiors of leaves. Here, chlorophyll absorbs solar energy in the red and blue parts of the spectrum. This energy is then used to split a water molecule into hydrogen and oxygen; in the process, the plants gain chemical energy that is used in a companion process that converts carbon dioxide into carbohydrates represented by CH_2O in the above equation.

The CO_2 used in this process gets into the interior of the leaf through small openings about 10 microns in diameter called stomata, which the plant can control like a valve, opening and closing to adjust the rate of transfer. The rate of CO_2 transfer is also a product of how much CO_2 is inside the leaf. For instance, if the interior of a leaf was filled with lots of CO_2, then even if the stomata were wide open, little new CO_2 could enter the leaf - it would be limited by how quickly the plant was consuming CO_2 in photosynthesis. Plants that are actively photosynthesizing generally have their stomata wide open, but there is sometimes a problem with this because with the stomata open, water from inside the plant escapes out into the air. As a consequence, when water is scarce, plants tend to close their stomata and thus they cannot photosynthesize at the maximum rate. Photosynthesis is also limited by the availability of other nutrients, especially nitrogen, which tends to be a limiting nutrient in many terrestrial ecosystems. However, as a general rule, if plants do have sufficient water, they will increase their rate of photosynthesis if there is more CO_2 in the air surrounding them - an effect known as CO_2-fertilization. An interesting thing happens when the rate of photosynthesis increases; the leaf interiors are filled with more CO_2, so the stomata close down to limit the intake of more CO_2, and in doing this, the plant also limits the amount of water that can escape. What this amounts to is an increased efficiency of water use by the plants - they grow more while using less water. But, there is an upper limit to this effect; at high levels of CO_2, plants do not respond to additional increases in atmospheric CO_2. Likewise, there is a lower limit to photosynthesis imposed by CO_2 levels, but we are not likely to approach such low levels in the next billion years or so (we will eventually have this problem since atmospheric CO_2 is slowly being drawn down through the precipitation of limestone).

Temperature is another important consideration in many life processes, and photosynthesis is no exception. As a general rule, the rates of most metabolic processes increase with temperature, but there is usually an upper limit where the high temperatures begin to destroy important enzymes, or otherwise inhibit life functions. For the majority of plants, this upper limit is not likely to come into play given the kinds of temperature changes we might expect in the space of a couple hundred years, so we can safely ignore it here (if our model does lead to temperature changes of greater than 10-20 °C in a hundred years, we would presume that there is some problem with the model as this is unrealistic behavior). Another important reason to take temperature into account in our photosynthesis flow is that it turns out that in most environments, an increase in temperature correlates with an increase in precipitation, and since many regions where plants grow are somewhat limited by water, especially towards the end of the growing season, increased precipitation leads to a greater yearly rate of photosynthetic uptake of atmospheric carbon.

How will we keep track of the changing temperature? Recall that an increase in atmospheric CO_2 leads to an increase in the greenhouse effect and thus a warmer planet. One option is that we could attach our model of the carbon cycle to one of our models for the climate system. Another, simpler option is to develop a simple relationship based on the known history of atmospheric CO_2 and global temperature. This has been done by a variety of people, and the result is an equation that looks like this:

$$\Delta T = 2.5 \times \frac{In\left(\frac{pCO2}{290}\right)}{In(2)} \,^{\circ}C$$

Where ΔT is the temperature difference, in °C, between the start and any time during the model simulation; ln is the natural log; pCO_2 is the concentration of CO_2 in the atmosphere in ppm (parts per million); 290 is the concentration (ppm) of CO_2 in the atmosphere around 1890, when our model is set to begin. In this way, we monitor the temperature relative to the starting temperature (which we'll call 14 °C). This logarithmic relationship means that for each doubling of CO_2, there is a 2.5 °C temperature increase.

Next, we must consider the problem of how to describe photosynthesis, globally, in the form of an equation that we can use in a model. We do get some help here from studies on a variety of different plants grown in greenhouses with varying levels of CO_2 and varying temperatures. We should realize, of course, that there are always uncertainties that arise when transferring results from the laboratory to the real world, and there are even more problems when you try to extrapolate to the scale of the whole world. The truth of the matter is that we don't know enough about the operation of our whole planet to solve this problem with confidence — the study of the global carbon cycle is in its early stages. And yet if we don't make some attempt to describe this process in the form of a global model, our understanding of the dynamics of the global carbon cycle will languish in the early stages. So we forge ahead.

The accumulated results indicate that our function for photosynthesis should satisfy the following characteristics:

- The present day global flow should be 120 Gt C/yr (one gigaton of carbon per year; 1Gt = 10^{15}g, also known as a petagram, Pg).

- The rate should increase at higher atmospheric CO_2 levels, but there should be a saturation point, where additional increases in atmospheric CO_2 yield no further increase in the rate of photosynthesis.
- The rate should increase with increasing temperature.

There are several ways to construct an equation that incorporates this information, but here, we use the following equations, adapted from Gifford:

$$F_p = \left(\frac{P_{max} \times pCO_{2\,eff}}{K_{hs} + PCO_{2\,eff}} \right) \times \left(1 + \left(Tsens_p \times \Delta T\right)\right)$$

$$P_{max} = \frac{\left(K_{hs} + \left(pCO_{2\,init} - pCO_{2\,min}\right)\right) \times F_{P\,init}}{\left(pCO_{2\,init} - pCO_{2\,min}\right)}$$

Where,

- F_p is the global rate of photosynthetic uptake of CO_2 in GtC/yr.
- P_{max} is a parameter with units of GtC/yr that is used to force the equation for F_p to give the proper value corresponding to the starting conditions of the model.
- K_{hs} = 72.5 ppm — the half-saturation value — the atmospheric CO_2 concentration at which the rate of photosynthesis, F_p, is half of the ultimate saturation value, given that particular temperature pCO_{2init} = 290 ppm — the pre-industrial (1880) atmospheric CO_2 concentration
- $pCO_{2\,min}$ = 30 ppm — value below which no photosynthesis can occur.
- $pCO_{2\,eff} = pCO_2 - pCO_{2\,min}$ — the effective atmospheric CO_2 concentration $F_{p,\,init}$ = 100 GtC/yr — the initial (pre-industrial) value for global photosynthesis.
- $Tsens_p$ = 0.04 — establishes the increase of F_p per degree of warming.
- $\Delta T = T - T_{init}$ — temperature difference relative to initial temperature. T_{init} is set to 0.

The figure shows what the equation for F_p looks like with varying levels of atmospheric CO_2 (and thus varying temperatures too); this kind of curve is called a rectangular hyperbola and is typical of many biological processes that have a saturation effect.

The K_{hs} parameter, the so-called half-saturation value, is the atmospheric CO_2 concentration that corresponds to a rate of photosynthesis (F_p) that is half the ultimate saturation value. K_{hs} is a key parameter in defining the shape of the curve, which is always constrained to give a photosynthesis rate of 100 Gt C/yr when the atmospheric CO_2 concentration is 290 ppm — the starting (~1890) value.

The temperature sensitivity part of the equation is simply a linear function, meaning that if you plot the temperature sensitivity versus the temperature, you get a straight line. The formulation here gives an increase in photosynthesis by a factor of 1.4 if the temperature increases 10°C. This is sometimes called the Q_{10} for the process, and a Q_{10} of 1.4 is in line with a variety of observations from experiments in greenhouses. It is important to note once again that there is an upper limit to this temperature sensitivity function in the real world.

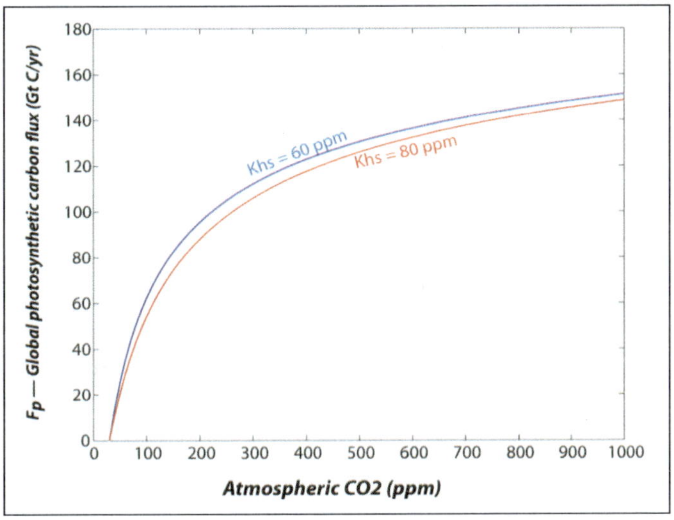

Defined in the manner outlined above, our photosynthesis flow is the same as what is commonly called Gross Primary Production. Gross Primary Production (GPP) differs from Net Primary Production (NPP), which is the gross minus respiration.

Plant Respiration

If we think of photosynthesis as the process of making fuel, then respiration can be thought of as the process of burning that fuel — using it for maintenance and growth. This process can be described in the form of a reaction, just like photosynthesis. That reaction is:

$$CH_2O + O_2 \rightarrow CO_2 + H_2O + \text{energy},$$

which is effectively the reverse of photosynthesis. Through respiration, plants (and animals) release water and carbon dioxide. They also release energy, in the form of heat, or infrared radiation. Thus, the processes of living organisms represent a kind of energy flow system; energy flows in and energy of another form flows out. Do the carbon flows involved in respiration and photosynthesis balance each other as the equations seem to imply? The answer is no — otherwise, how could organisms grow.

Experiments on a variety of plants indicate that the ratio of photosynthesis to respiration is generally about 2 to 1, and this ratio does not appear to depend on the temperature. Or more precisely, the respiration does depend upon temperature, but in approximately the same way that photosynthesis does, so the ratio between them stays the same. When plants are young, and growing rapidly, but with not much biomass to maintain, this ratio is even higher; in older, larger plants, this ratio is lower since more carbon needs to go towards maintenance. For our purposes, we will assume that the global collection of plants has a ratio of about 2 to 1, so we will formulate plant respiration in the following manner:

$$F_{pr} = F_p{}^{*}(50/100)$$

where, F_{pr} is the rate of global plant respiration in Gt C/yr and F_p is the rate of global photosynthesis by land plants.

Litter Fall and Below-ground Addition to the Soil

Dead plant material enters the soil in two ways -- it falls on the surface as litter, and it is contributed below the surface from roots. The relative importance of these two pathways into the soil carbon reservoir vary according to the plants in an ecosystem, but it appears that the two are commonly about equal, which may seem a bit surprising since loss of organic carbon from root systems is a process that we generally don't see. In our model, we will lump these processes together and call them litter fall, keeping in mind that this is only half of the real story (later, we can enhance our model by breaking the soil up into several different boxes, and these two flows will no longer be lumped together). Litter fall is set to be the difference between the photosynthetic uptake of carbon and the return of carbon through plant respiration. If this were not the case, then the size of the land biota reservoir would be growing or declining, which may in fact be the case (it's doing both in different parts of the Earth), but we would like our model to be more or less in a steady state to begin with. In reality, it is the litter fall that is actually measured in studies of carbon flow through ecosystems. Litter fall combined with a measure of the gross primary productivity (the total amount of carbon used in photosynthesis) gives an estimate of the plant respiration flow according to the following equation:

Gross Primary Productivity = Plant Respiration + Litter Fall

Having already chosen the initial rates for photosynthesis and plant respiration, at 100 and 50, this leaves us with a value of 60 Gt C/yr for the rate of carbon added to the soil reservoir through the process of litter fall. But, we don't want this to be a constant value; it will undoubtedly change as a function of the size of the land biota reservoir. So, we can define this flow in the form of a standard draining flow:

F_{lf} = land_biota * (50/INIT(land_biota)

where, F_{lf} is the flow in Gt C/yr and land_biota is the amount of carbon stored in the land biota (plants) at any given time.

Soil Respiration

Respiration (sometimes called decay) also occurs within the soil, as microorganisms consume the dead plant material. In terms of a chemical formula, this process is the same as described above for plant respiration, only in this case, the microbes do not make the fuel themselves.

There is an unseen but fascinating universe of microbes living within the soil and they are the key means by which nutrients such as carbon and nitrogen are cycled through the soil system. A great diversity of microorganisms lives in the soil — perhaps as many as 1000 species in a cubic centimeter — and they are capable of consuming tremendous quantities of organic material. Much of the organic material added to the litter (the accumulated material at the surface of the soil) or within the root zone each year is almost completely consumed by microbes; thus there is a reservoir of carbon with a very fast turnover time — on the order of 1 to 3 years in many cases. The by-products of this microbial consumption are CO_2, H_2O, and a variety of other compounds, collectively known as humus. Humus is a much less palatable compound, as far as microbes are concerned, and is not decomposed very quickly. After it is produced at shallow levels within the soil, it generally moves downward and accumulates in regions of the soil with high clay content. Part of the reason it accumulates in the lower parts of the soil is that there tends to be less oxygen in that environment, and the lack of oxygen makes it even more difficult for microbes to work on this humus and decompose it further. But eventually, due to various processes that stir the soil, this humus moves back up to where there is more oxygen and then the microbes will eventually destroy the humus and release some more CO_2. This humus then constitutes another, longer-lived reservoir of carbon in the soil. ^{14}C dates on some of this soil humus give ages of several hundred to a thousand years old. Taken together, the fast and slow decomposition processes, both driven by microbes, lead to an average residence time of around 20 to 30 years for most soils. The data used in our global carbon cycle model lead to a residence time of about 26 years for the global soil carbon reservoir.

These microbes (considered in terms of their respiratory output) are very sensitive to the organic carbon content of the soil as well as the temperature and water content, respiring faster at higher carbon concentrations, higher temperatures and in moister conditions (although if the soil is flooded with water, conditions are worse since no oxygen can get into the soil - the majority of the microbes need oxygen to respire, as can be seen from the general equation for respiration given under the discussion of plant respiration). Studies seem to indicate that the soil respiration is in fact even more sensitive to temperature than photosynthesis and plant respiration. In general, higher temperatures tend to correspond with higher rates of precipitation, so we can consider the effects of water to go hand in hand with temperature. In our model of the carbon cycle, we will use an expression for soil respiration that takes these observations into account:

$$F_{sr} = Soil \times \left(\frac{49.4}{INIT(Soil)} \right) \times \left(1 + \left(Tsens_{sr} \times \Delta T\right)\right)$$

$$Tsens_{sr} = 0.10 \text{ to } 0.01$$

The temperature sensitivity part of the equation is a linear function like that used in defining photosynthesis. The $Tsens_{sr}$ value of 0.1 gives an increase in soil respiration by a factor of 2.0 if the temperature increases 10°C. This gives a Q_{10} of 2.0 and is in line with a variety of observations (which generally are not in close agreement; this is not a well constrained part of the model). In our initial model, we'll set this to 0.03, because it gives a better match with the observed record of atmospheric CO_2. It is important to note once again that there is an upper limit to this temperature sensitivity function in the real world, but we do not expect to approach it in this modeling exercise.

Permafrost — An Unknown

In recent years, increasing attention has been directed at permafrost soil carbon since the Polar Regions are warming much faster than the rest of the globe. As the permafrost melts, carbon that was added to these soils by processes like litter fall, will become available for soil microbes to respire and release to the atmosphere. In fact, it is almost surely happening already, but given that much of the permafrost is still frozen, we have probably not seen the real manifestation of this source of carbon. Estimates are variable, but a figure like 1000 to 1500 Gt of carbon reflects the current thinking; this is a huge amount of carbon and has the potential to significantly alter the future of atmospheric CO_2 levels. As the permafrost begins to melt, some estimates are that it will contribute something in the range of 2-5 Gt C/yr, which is large, compared to the human-related changes. Of course, some of this released carbon will be offset by new carbon sequestered into these formerly frozen soils, but initially, the system will not be in equilibrium and these regions can be expected to be a net source of CO_2 to our atmosphere.

This process could be added to the model by first creating a new reservoir called permafrost C and then adding a flow from it to atmosphere. The flow, which might be called melting could be defined as follows:

$$F_{pf} = .05 * \left(\frac{\text{PermafrostC}}{\text{INIT(PermafrostC)}} \right) * \left(1 + \left(\text{Tsens}_{pf} \times \Delta T\right)\right)$$

$$\text{Tsens}_{pf} = 40$$

Here, the initial flow value is set to be very small, such that at the start of the model, this is an insignificant flow. The temperature sensitivity here is very large, but this creates a flow of around 3-4 Gt C/yr with the kinds of warming we expect by the end of the century.

Runoff

Although most of the carbon loss from the soil reservoir occurs through respiration, some carbon is transported away by water running off over the soil surface. This runoff is eventually transported to the oceans by rivers. The actual magnitude of this flow is a bit uncertain, although it does appear to be quite small. The most recent estimates place it at 0.6 Gt C/yr; we'll define this as a standard draining process (first-order kinetic process) as follows:

$$F_r = Soil \times \left(\frac{0.6}{INIT(Soil)} \right)$$

where, F_r is the runoff flow to the ocean in GtC/yr.

Processes of Carbon Flow in the Oceanic Realm

Ocean - Atmosphere Exchange

Carbon dioxide can be dissolved in seawater, just as it can be dissolved in a can of soda; it can also be released from seawater as the CO_2 from a soda can also be released. This transfer of gas back and forth between a liquid and the atmosphere is an extremely important process in the global carbon cycle since the oceans are such an enormous reservoir with the potential to store and release significant quantities of CO_2. In general, this flow depends on the concentration of CO_2 in the atmosphere and the oceans, but it gets a little complicated because the concentration of CO_2 in seawater depends on a number of other factors. To understand this process, we need to have some sense of what happens to CO_2 once it gets dissolved in seawater.

Carbonate Chemistry of Seawater

When CO_2 enters seawater, it reacts with water and forms a series of products, as described in the following equation:

$$CO_2 + H_2O \rightarrow H_2CO_3 \rightarrow H^+ + HCO_3^{-1} \rightarrow 2H^+ + CO_3^{-2}$$

This means that in seawater, you can find all of these different forms or species of carbon co-existing, but in reality, bicarbonate (HCO_3^-) is the dominant form of inorganic carbon; carbonate (CO_3^{2-}) and dissolved CO_2 are important, but secondary.

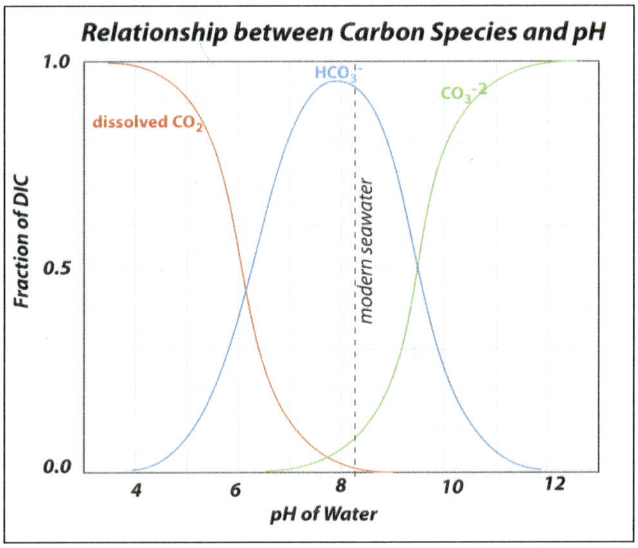

The relative proportions of the different carbon species depends on the alkalinity and the total DIC; these proportions in turn determine the pH. Salinity and temperature also exert some control on the proportions of carbon species. In the oceans today, most of the carbon is in the form of HCO_3^- with relatively small amount of CO_3^{2-} and dissolved CO_2.

In the above equation, the double-headed arrows mean that the reactions can go in both directions and generally do until some balance of the different compounds is achieved -- a chemical equilibrium. Each of these reactions has associated with it an equilibrium constant, which establishes the relative concentrations of the compounds on either side of the reaction. For instance, if we isolate two of the above reactions,

$$H_2CO_3 \rightarrow H^+HCO_3^{-1}$$
$$HCO_3^{-1} \rightarrow H^+ + CO_3^{-2}$$

we can say that the equilibrium constants, k_1 and k_2, are given by a ratio formed by the concentrations of the various compounds involved:

$$k_1 = \frac{[H][HCO_3]}{[H_2CO_3]}, \quad k_2 \frac{[H][CO_3]}{[HCO_3]}.$$

We can do some algebra with these equations, just like normal equations. Our goal is to find an expression for the concentration of H_2CO_3, so we begin by rearranging the equation on the left

$\left(k_1 = \frac{[H][HCO_3]}{[H_2CO_3]},\ k_2 \frac{[H][CO_3]}{[HCO_3]}\right)$ to the following form:

$$[H_2CO_3] = \frac{[H][HCO_3]}{k_1}$$

Then we rearrange the right side of $\left(k_1 = \frac{[H][HCO_3]}{[H_2CO_3]},\ k_2 \frac{[H][CO_3]}{[HCO_3]}\right)$ so that it becomes:

$$[H] = \frac{k_2[HCO_3]}{[CO_3]}$$

If we know or calculate the concentration (or activity) of hydrogen — [H] — it is an easy step to get the pH of the seawater, just take the negative log of [H], which our model can keep track of. Next, we substitute $\left([H] = \frac{k_2[HCO_3]}{[CO_3]}\right)$ into $\left([H_2CO_3] = \frac{[H][HCO_3]}{k_1}\right)$, to give us our desired equation expressing H_2CO_3 in terms of HCO_3^- and CO_3^{2-}.

$$[H_2CO_3] = \frac{k_2}{k_1}\left(\frac{[HCO_3]^2}{[CO_3]}\right)$$

Ultimately, we want to get an expression for the concentration of CO_2 gas contained in the seawater at equilibrium; this is usually expressed as the partial pressure of CO_2, with units of μatm (micro-atmospheres), or ppm (parts per million, by volume) rather than a typical concentration, which would have units of moles per cubic meter of water. The partial pressure is given by the following formula:

$$pCO_2 = K_3[H_2CO_3]$$

in which K_3 is a slightly different kind of equilibrium constant that incorporates the solubility of CO_2 in seawater, which is a function of temperature and salinity. Our next step is to combine the various equilibrium constants into a single value as:

$$KCO_2 = K_3 \times \frac{k_2}{k_1}$$

keeping in mind that this value will be a function of temperature (also salinity). Then we can substitute ($[H_2CO_3]=\frac{k_2}{k_1}\left(\frac{[HCO_3]^2}{[CO_3]}\right)$) and ($KCO_2=K_3\times\frac{k_2}{k_1}$) into ($pCO_2=K_3[H_2CO_3]$) to obtain another expression for the partial pressure of carbon dioxide gas in seawater:

$$pCO_2=KCO_2\times\left(\frac{[HCO_3]^2}{[CO_3]}\right)$$

The next thing we need to do is to find expressions for the concentrations of carbonate and bicarbonate in terms of the total amount of carbon dissolved in seawater, which will change over time as the oceans release or absorb CO_2 from the atmosphere. First, we need to define a term for the total concentration of inorganic carbon in solution:

$$\sum CO_2=[HCO_3]+[CO_3]$$

This is an approximation since it ignores CO_2 gas and H_2CO_3, but both of these are very minor components. This sum, the concentration of total dissolved carbon, is simply equal to the amount of carbon in the ocean reservoir divided by the volume of the ocean. Looking at the above equation, we see that it does not tell us the relative proportions of the carbonate and bicarbonate — just their combined concentration. So, we need to find some other means of establishing the proportions of these two forms of carbon. We get some help from the fact that the relative proportions of bicarbonate (HCO_3^-) and carbonate (CO_3^{2-}) play an important role in establishing the balance of positive and negative charges in seawater. Ions like calcium (Ca^{2+}), sodium (Na^+), , potassium (K^+), and magnesium (Mg^{2+}) all add positive charges to seawater; these are partially countered by chloride (Cl^-), sulfate (SO_4^{2-}), nitrate (NO_3^-), and bromide (Br-) ions, but the result is a slight deficit of negatively charged ions. This difference is essentially made up by the carbonate and bicarbonate ions. If more negative charge is needed, then more of the carbon occurs in the form of carbonate since it has a charge of minus 2, while if less negative charge is needed, more of the carbon exists in the form of bicarbonate. The excess positive charge of seawater that needs to be balanced by the different forms of dissolved carbon is called the alkalinity of the seawater. The alkalinity is defined as:

$$Alk=[HCO_3]+2[CO_3]$$

Note that the carbonate ion counts twice since it has a minus 2 charge. It is important to realize that the total alkalinity is really determined by the other ions in solution — the ones mentioned above. This means that for the purposes of our model, we can keep the alkalinity constant, although we could define slightly different alkalinities for the warm and cold surface waters of the oceans if we made a more elaborate model. Looking at the above equation, you can see that with a given alkalinity, if we have just a little dissolved carbon, more of the carbon will be in the form of the carbonate ion (CO_3^{2-}) in order to make up the charge imbalance, but if we have a high concentration of dissolved carbon, there will be a greater proportion of the bicarbonate ion (HCO_3). The concentrations of both the carbonate and bicarbonate ions can be expressed as a function of both the alkalinity and the concentration of total dissolved inorganic carbon, ΣCO2, as shown in the following:

$$[CO_3]=Alk-\sum CO_2$$
$$[HCO_3]=2\sum CO_2-Alk$$

Here, we need to remember that this is an approximation because we ignored the term for H_2CO_3 in equation $\sum CO_2 = [HCO_3] + [CO_3]$. In the model we will eventually experiment with, we use a more precise formulation that does not ignore the concentration of H_2CO_3. This increases the complexity of the algebra, giving us a quadratic equation whose solution ends up as:

$$[HCO_3] = \frac{\sum CO_2 \sqrt{\sum CO_2 - Alk\left(2\sum CO_2 - Alk\right) \times \left(1 - 4\frac{k_2}{k_1}\right)}}{\left(1 - 4\frac{k_2}{k_1}\right)}$$

$$[CO_3] = \left(\frac{Alk - [HCO_3]}{2}\right)$$

Regardless of whether we consider the more digestible form:

$$\left([CO_3] = Alk - \sum CO_2 \text{ and } [HCO_3] = 2\sum CO_2 - Alk\right)$$

or

the more precise form of expressing HCO_3^- and CO_3^{2-}:

$$\left([HCO_3] = \frac{\sum CO_2 \sqrt{\sum CO_2 - Alk\left(2\sum CO_2 - Alk\right) \times \left(1 - 4\frac{k_2}{k_1}\right)}}{\left(1 - 4\frac{k_2}{k_1}\right)} \text{ and } [CO_3] = \left(\frac{Alk - [HCO_3]}{2}\right)\right),$$

we are finally set, because if we look at the equation for the partial pressure of CO_2,

$$pCO_2 = KCO_2 \times \left(\frac{[HCO_3]^2}{CO_3}\right)$$

we see that all of the terms can be expressed in terms of components of the model. We have reached the light at the end of the algebraic tunnel.

As can be seen, the chemistry of carbon in seawater is relatively complex, but it turns out to be extremely important in governing the way the global carbon cycle operates and explains why the oceans can swallow up so much atmospheric CO_2 without having its own CO_2 concentration raise very much.

Ocean - Atmosphere Exchange Revisited

The exchange of gas between the air and the sea is, in theory, governed by the rate of diffusion and the distance the diffusion has to cover, as indicated the figure.

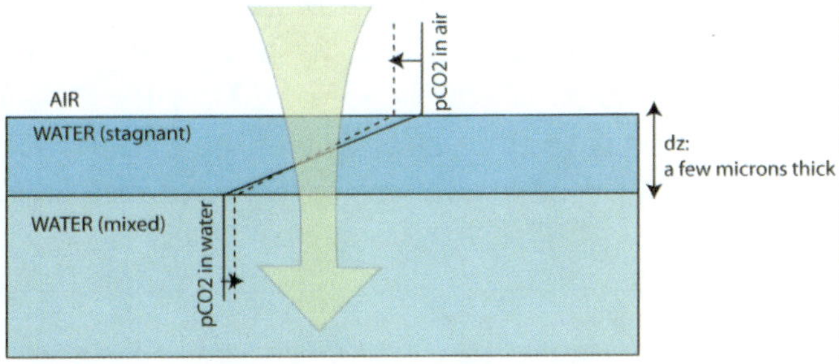

Here, the atmosphere has a higher concentration of CO2 than the ocean, so it will flow into the ocean at a rate governed by the diffusion constant and the concentration gradient:

$$\frac{dC}{dt} = D \times \left(\frac{C_{atm} - C_{oc}}{dz} \right)$$

If the concentration gradient goes the other way — higher in the water than the air — the flow of CO2 goes up into the air.

Air-Sea Carbon Dioxide Exchange.

As depicted in the figure above, the concentrations in the air and the sea are relatively constant (spatially, not temporally) since these two media undergo rapid and turbulent mixing that would tend to even out any systematic variations. The exception to this is a thin layer of water, just 20 to 40 microns thick, which, because of the surface tension of the water, is unable to mix well. This stagnant film is the barrier across which the diffusion has to occur. The rate of gas transfer, or flux, is given by the equation shown in the figure. The units of the diffusion coefficient are in m^2 /yr and the units of the concentrations would be given in units like g/m^3, so the overall units are in g/m^2 yr. When this quantity is multiplied by the area of the ocean, then we have the overall rate of transfer in g/yr. In our model, we will represent this process in the following way:

$$F_{ao} = k_{ao}\left(pCO_{2\,atm} - pCO_{2\,OC}\right)$$
$$k_{ao} = 0.278\left(Gt\ C\ yr^{-1}\ ppm^{-1}\right)$$

The constant we use here, k_{ao}, combines the diffusion constant, the stagnant film thickness and the area of the oceans. We will take this to be a constant value, but it is important to consider that it includes incorporates the stagnant film thickness, which is related to wind velocity. Higher velocity winds will lead to a thinner stagnant film and thus the gas transfer will be faster.

In the real world, there are important variations in the gas transfer between the ocean and atmosphere, which can be seen in the figure below, which represents a kind of snapshot of this transfer across the globe. The units here are grams of C per m^2 per year and each box is about 1e6 m^2. The red, orange, and yellow colors represent places where the oceans are giving up CO_2 to the atmosphere; the blue and purple areas are places where the oceans are sucking up atmospheric CO_2. Summing these up, we find that the oceans are taking up around 92 Gt C/yr and they are releasing about 90 Gt C/yr — for a net flow of 2 Gt C/yr into the oceans. This exchange is variable in space and time, but a few general features can be pointed out. In general, the colder parts of the oceans absorb CO2and the warmer parts release CO_2 into the atmosphere. This makes sense because CO_2 is more soluble in colder water another way to say this is that the equilibrium constants in the carbonate chemistry reactions described above are sensitive to temperature such that at colder temperatures, more of the carbon is in the other carbon species, so the pCO_2 of the ocean water is lower when the temperature is colder. One exception is the ocean around Antarctica, where the water is cold and yet it releases a good deal of CO_2 to the atmosphere; the reason for this is that there is upwelling of deep water in that zone and the deep ocean has much higher pCO_2 values than the surface waters, largely due to the effects of the biological pump.

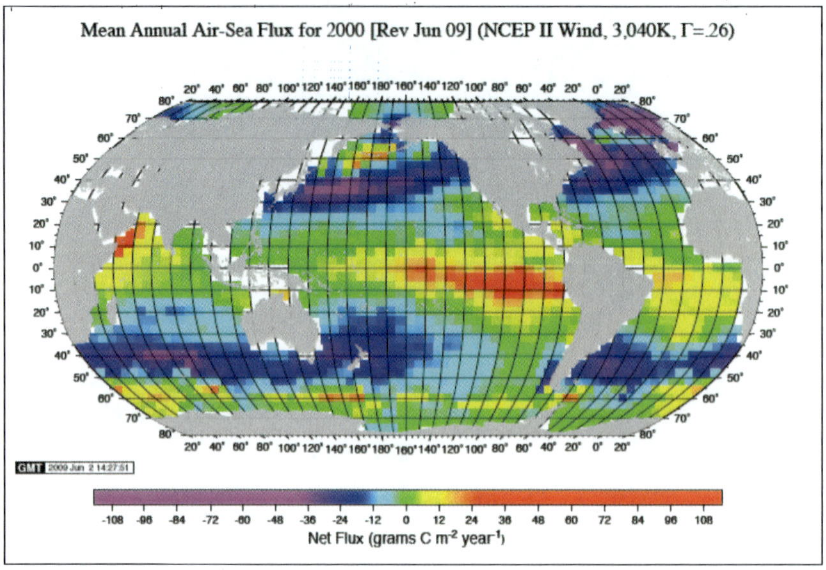

Marine Biota Exchange — Biologic Pump

The surface waters of the world's oceans are home to a great number of organisms that include photosynthesizing phytoplankton at the base of the food chain. These plants (and cyanobacteria) utilize CO_2 gas dissolved in seawater and turn it into organic matter, and just like land plants, these phytoplankton also respire, returning CO_2 to the surface waters.

At the same time, many planktonic organisms extract dissolved carbonate ions from seawater and turn them into $CaCO_3$ shells. When these planktonic organisms die, their soft parts are mainly consumed and decomposed very quickly, before they can settle out into the deeper waters of the oceans. This decomposition thus returns carbon, in the form of CO_2, to seawater. The net transfer then is defined in our model as follows:

$$F_{ab} = 10 * \left(\frac{Ocean\ Biota}{INIT\left(Ocean\ Biota\right)} \right)$$

However, some of the organic remains and the inorganic calcium carbonate shells will sink down into the deep oceans, thus transferring carbon from the shallow surface waters into the huge reservoir of the deep oceans. This transfer is often referred to as the biologic pump, and it causes the concentration of CO_2 gas, and also DIC (ΣCO_2), the concentration of total dissolved inorganic carbon in the surface waters to be less than that of the deeper waters. This can be seen in figure, which shows the vertical distribution of ΣCO_2 (and also alkalinity) in a profile view for some of the major regions of the world's oceans.

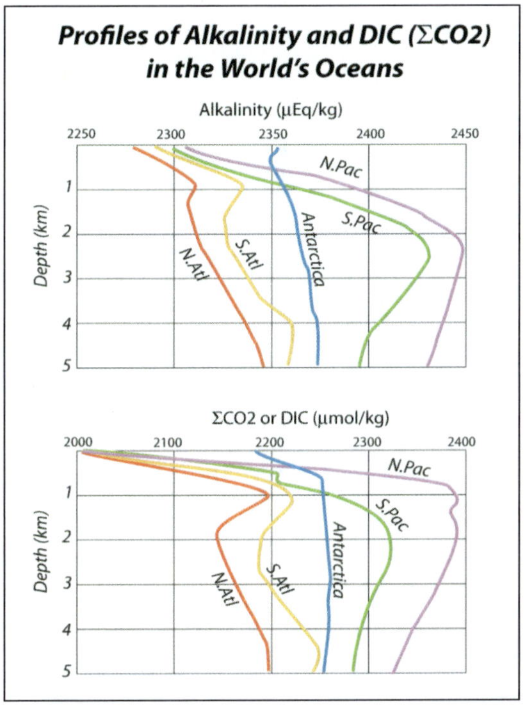

Figure representative vertical profiles of alkalinity and concentration of total inorganic carbon (DIC or ΣCO_2) from the world's oceans. In general, the surface waters are depleted in alkalinity and DIC because of the biologic pump, which exports carbon (both inorganic and organic) from the surface waters to the deep ocean. The carbonate portions of the marine biota dissolve in the deep waters and release Ca2+ ions, thereby increasing the alkalinity of the deep ocean, and at the same time, depleting the alkalinity of the surface ocean.

Why is the alkalinity reduced in the surface waters? For the same reason that DIC (ΣCO_2) is depleted. Planktonic organisms make shells of $CaCO_3$, and when these sink to the seafloor, they carry Ca^{2+} ions with them, thus reducing the alkalinity. Much of this $CaCO_3$ is later dissolved when it reaches deeper parts of the oceans, which explains the higher alkalinity values in the deep waters, as seen in figure. By controlling the concentration of CO_2 gas dissolved in the surface waters, the planktonic organisms exert a strong influence on the concentration of CO_2 in the atmosphere. For instance, if the biologic pump were turned off, atmospheric CO_2 would rise to about 500 ppm (relative to the current 350 ppm); if the pump were operating at maximum strength (i.e., complete utilization of nutrients), atmospheric CO_2 would drop to a low of 140 ppm. Clearly, this biologic pump is an important process. In our model, the pump is defined as follows:

$$F_{bp} = 10 \times \left(\frac{Ocean\ Biota}{INIT\left(Ocean\ Biota\right)} \right)$$

What controls the strength of this biologic pump? The photosynthesizing plankton requires nutrients in addition to CO_2 in order to thrive; specifically, they require nitrogen and phosphorus. Most of these plants need P, N, and C in a ratio of 1:16:125 and since at present, the ratio of P to N in ocean water is about 1:16, both P and N limit the growth of this phytoplankton. Photosynthetic activity of plankton can be mapped out by satellites tuned to record differences in water color due to the presence of chlorophyll. This distribution is shown in figure.

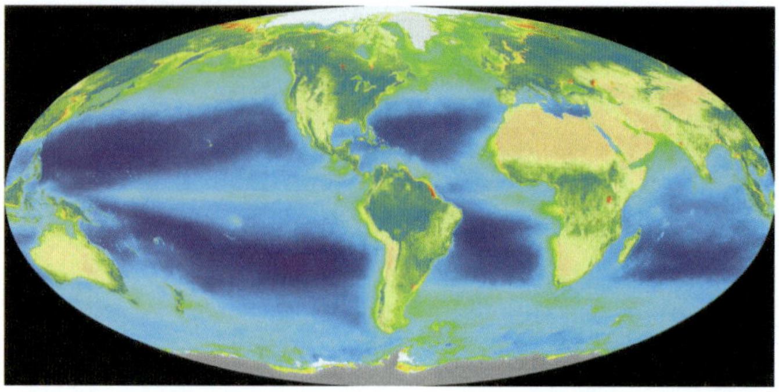

If the nutrients in seawater were being utilized to the maximum extent possible, there would be practically no P and N dissolved in seawater, but in fact, as shown by the figure below, the concentration of P tells us that the biologic pump is not operating at maximum efficiency. In the map below, the purple regions represent regions with no phosphate in the surface water of the oceans, meaning that there is simply a lack of nutrients, or that all the nutrients are utilized. In particular, it is the cold, Polar Regions that are not utilizing all of the available nutrients. This may be due in part to the temperature, but it may also be related to a paucity of iron, a minor nutrient that is apparently lacking in the colder regions, especially in the southern ocean ringing Antarctica.

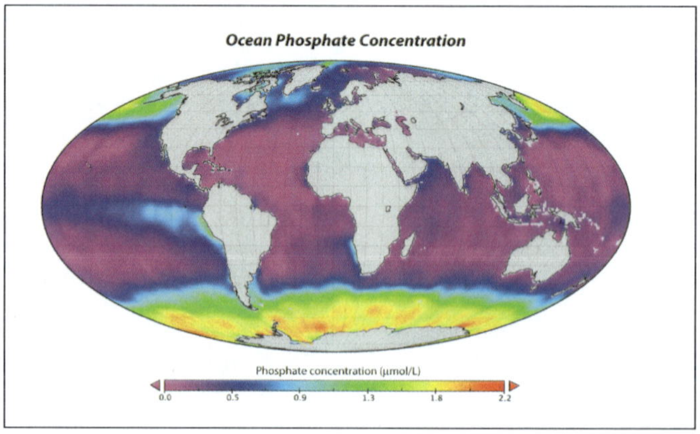

We might wonder whether or not it is likely that global warming might lead to a greater utilization of nutrients by the marine biota. If so, this would lead to a more vigorous biologic pump in the oceans that might tend to limit the increase of atmospheric CO_2. It seems plausible given the observation that in today's ocean, there is greater nutrient utilization in areas where the ocean is warmer. Is there a record of this in the past. Since it is now much warmer than it was during the last glaciation, we might investigate whether the biological pump was any weaker then relative to the present. Not surprisingly, people have looked into this question and the data indicate that the pump was, if anything, slightly stronger during the last glaciation. How is this known? It is based on record of the ratio of different isotopes of carbon, preserved in the shells of plankton that were deposited on the seafloor. Intense biological productivity in the surface waters alters - relative to the rest of the ocean - the carbon isotope ratio because organisms have a strong preference for one isotope of carbon (^{12}C) than another (^{13}C). The magnitude of this isotope shift is then a function of how intense the biological pump has been.

Downwelling

Downwelling transfers cold surface waters into the deep interior of the oceans, and as a result, carbon is transferred as well. The magnitude of the flow is thus a function of the volume of water flowing and the average concentration of carbon in the cold surface waters, which is itself a function of the total amount of carbon stored in this reservoir, assuming that the size of the reservoir is not changing appreciably over the few hundred years that this model is intended to be used for.

Downwelling occurs primarily near the poles, where surface waters are strongly cooled by contact with the air. This cooling leads to a density increase. The formation of ice from seawater at the margins of Antarctica increases the salinity of the seawater there, adding to the density of the water. This dense water then sinks and flows through the deep oceans; effectively mixing them on a timescale of about 1000 years or so (the Atlantic Ocean mixes somewhat faster, which helps explain the smaller ΣCO_2 and alkalinity gradients seen in figure). In our model, the flow of carbon from the cold, shallow water to deep water will simply be a function of how much carbon is stored in the cold surface water reservoir. This implies that we are not altering the volumetric flow of downwelling water, which is probably acceptable if we are only concerned with a few hundred years. In the form of an equation, this flow is:

$$F_{dwn} = 90.6 * \left(\frac{Surf\ Ocean}{INIT(Surf\ Ocean)} \right)$$

where, F_{dwn} is the flow of carbon due to downwelling, in Gt C/yr. This equation is set so that at the beginning of the model, the magnitude of this flow will be 90.6, which is within the range of current estimates, adjusted here to make the system be in a steady state.

Upwelling

Upwelling is the converse of downwelling, and as the deep waters rise to the surface, they bring with them carbon. The total transfer of carbon is thus a function of the volume of water involved in this flow and the amount of carbon stored in the deep ocean reservoir. Upwelling occurs in areas of the oceans where winds and surface currents diverge, moving the surface waters away from a region; in response, deep waters rise up to fill the "void". Upwelling occurs along the equator, where there is a strong divergence, and also along the margins of some continents, such as the west coast of South America. This upwelling water also brings with it nutrients such as nitrogen and phosphorus, making these waters highly productive.

We will define the flow of carbon resulting from upwelling in a like manner to the downwelling flow:

$$F_{up} = 100 * \left(\frac{Deep\ Ocean}{INIT(Deep\ Ocean)} \right)$$

Note that the amount of carbon transferred by this flow is greater than the downwelling flow. This is not because the volume of flow is different in these two processes, but rather because the concentration of carbon in the deep waters of the ocean is greater than that in the shallow surface waters, due in part to the operation of the biologic pump mentioned above.

To make our model simpler, we combine the upwelling and downwelling flows into one flow called ocean mixing, which is defined as follows:

$$F_{omx} = F_{up} - F_{dwn}$$

The flow is defined as a biflow, meaning that depending on the sign of this difference, carbon can move either way.

Sedimentation

Some of the carbon, both organic and inorganic (i.e., calcium carbonate shells) produced by marine biota and transferred to the deep oceans settles out onto the sea floor and accumulates there, eventually forming sedimentary rocks. The magnitude of this flow is small -- about 0.6 Gt C/yr -- relative to the total amount transferred by sinking from the surface waters -- 10 Gt C/yr. The reason for this difference is primarily because the deep waters of the oceans dissolve calcium carbonate shell materials; below about 4 km, the water is so corrosive that virtually no calcium carbonate material can accumulate on the sea floor. In addition, some of the organic carbon is consumed by organisms living in the deep waters and within the sedimentary material lining the sea floor. This consumption results in the release of CO_2 into the bottom waters and thus decreases the amount of carbon that is removed from the ocean through this process. It is worth noting that the process of organic carbon consumption on the seafloor is another microbial process and is very similar to the soil respiration flow described above. Since the microbes living on the seafloor require oxygen to accomplish this task, the supply of oxygen to the seafloor by deep currents is an important part of this process.

We will define the sedimentation flow to be a fixed percentage of the amount of carbon transferred to the deep waters by the biologic pump flows from the shallow marine biota reservoirs. The equation for this flow is thus:

$$F_{sed} = 0.6 \times \left(\frac{Bio\ Pump}{10} \right)$$

where, F_{sed} is the transfer of carbon from the deep ocean reservoir to the sedimentary rock reservoir, and BioPump the flow of carbon from the surface ocean biota. This equation is set up so that initially, with BioPump set to total 10 Gt C/yr, the F_{sed} flow will have a value of 0.6 Gt C/yr.

Volcanism and Metamorphism

When sedimentary rocks deposited on oceanic crust are subducted, they may melt or undergo metamorphism; in either case, the carbon stored in calcium carbonate - limestone - is liberated in the form of CO_2, which ultimately is released at the surface. The CO_2 may come out when a volcano erupts, or it may slowly diffuse out from the interior via hot springs, but in both cases, it represents a transfer of carbon from the reservoir of sedimentary rocks to the atmosphere. The magnitude of this flow is quite small, and is adjusted here to a value of 0.6 Gt C/yr in order to create a model in steady state. This flow is defined as a constant in the model, although in reality, it will vary according to the timing of large volcanic eruptions. An extremely large volcanic eruption may emit carbon at a rate of around 0.2 Gt C/yr for a year or two, creating a minor fluctuation.

Processes of Carbon Flow in the Human Realm

Fossil Fuel Burning

Another pathway for carbon to move from the sedimentary rock reservoir to the atmosphere is through the burning of fossil fuels by humans. Fossil fuels include petroleum, natural gas, and coal, all of which are produced by slow transformation of organic carbon deposited in sedimentary rocks - essentially the fossilized remains of marine and land plants. In general, this transformation takes many millions of years; most of the oil and gas we now extract from sedimentary rocks is on the order of 70-100 million years old. These fossil fuels are primarily composed of carbon and hydrogen. For instance, methane, the main component of natural gas, has a chemical formula of CH_4; petroleum can be represented by the formula of CH_2. The combustion of fossils fuels involves the use of oxygen and the release of carbon dioxide and water, as represented by the following description of burning petroleum:

$$2CH_2 + 3O_2 \rightarrow 2CO_2 + 2H_2O$$

Beginning with the onset of the industrial revolution at the end of the last century, humans have been burning increasing quantities of fossil fuels as our primary energy source. As a consequence, the amount of CO_2 emitted from this burning has undergone an exponential rise that follows the exponential rise in the human population. The magnitude of this flow is currently about 5 to 6 Gt C/yr. This number also includes the CO_2 generated in the production of cement, where limestone is burned, liberating CO_2.

This flow has changed considerably over time, as human population has increased and as our economies have become more industrialized with a big thirst for the energy provided from combustion of fossil fuels.

Land-use Changes - Forest Burning and Soil Disruption

The other form of human alteration of the global carbon cycle is through forest cutting and burning and the disruption of soils associated with agriculture. When deforestation occurs, most of the plant matter is either left to decompose on the ground or it is burned, the latter being the more common occurrence. This process reduces the size (the mass) of the land biota reservoir and the burning adds carbon to the atmosphere. Land-use changes other than deforestation can also add carbon to the atmosphere. Agriculture, for instance, involves tilling the soil, which leads to very rapid decomposition and oxidation of soil organic matter. This means that in terms of a system, we are talking about two separate flows here — one draining the land biota reservoir, the other draining the soil reservoir; both flows transfer carbon to the atmosphere. Current estimates place the total addition to the atmosphere from forest burning and soil disruption at around 1.5 Gt C/yr; estimates divide this into 70% to 50% forest burning, with soil disruption making up the remainder. The actual history of this alteration to the natural carbon cycle is not well-constrained — not nearly as well known as the fossil fuel burning history — but we include a reasonable history that reflects patterns of land settlement and forest clearing.

Oxygen Cycle

The oxygen cycle elaborates how oxygen circulates in various forms through nature. Oxygen occurs freely in the air, trapped in the earth crust as chemical compounds, or dissolved in water. Oxygen in the atmosphere is about 21%, and it is the second most abundant gas after nitrogen. It is mostly utilized by living organisms, especially man and animals in respiration. Oxygen is also the most common element of human body.

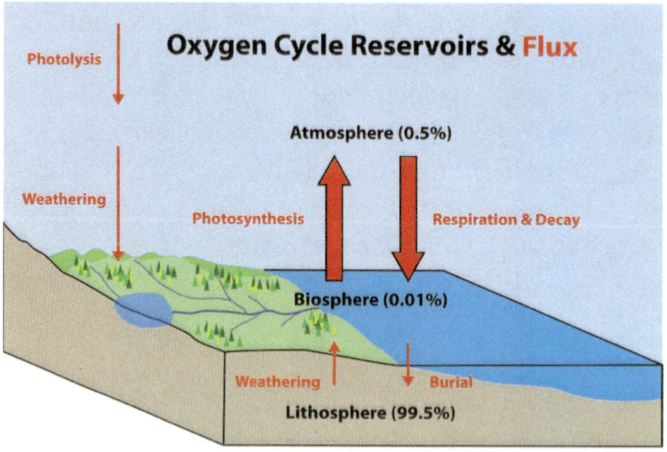

Oxygen is also used during combustion, decomposition, and oxidation. The circulation of oxygen is through three main flow systems including the (air) atmosphere, the biosphere, and the earth's crust. In the oxygen cycle, the main driving factor is photosynthesis which is the process whereby green plants and algae make their own food by use of solar energy, water, and carbon dioxide to gives off oxygen as a by-product.

Hence, for oxygen to remain in the atmosphere, it has to circulate through various forms of nature which is essentially termed as the oxygen cycle. The circulation depends on the various activities on Earth.

Process of Oxygen Cycle

Atmosphere (Air)

The atmosphere carries a small quantity of all oxygen, only about 0.35% of the entire earth's oxygen. In the atmosphere, oxygen is released by the process known as photolysis. Photolysis happens when the ultraviolet radiation of sunlight breaks apart oxygen-containing molecules such as nitrous oxide and atmospheric water to release free oxygen. The surplus oxygen recombines with other oxygen molecules to form ozone while the rest is freed into the atmosphere. Ozone is the layer that helps to shield the Earth from the dangerous ultra violet rays.

Biosphere

The biosphere carries the smallest quantity of all earth's oxygen, about 0.01%. In the biosphere, the major oxygen cycles are photosynthesis and respiration. In these two processes of the oxygen cycle, it is interconnected with the carbon cycle and the water cycle. During photosynthesis, plants and planktons use sunlight energy, water, and carbon dioxide to make food (carbohydrates) and release oxygen as a by-product. As such, plants and planktons are the main producers of oxygen in the ecosystem. They take in carbon dioxide and give out oxygen. Plants are estimated to replace about 99% of all the oxygen used.

On the other hand, respiration happens when humans and animals breathe in oxygen which is used during metabolism to break down carbohydrates and exhale carbon dioxide as a by-product. Such free carbon dioxide is then released into the environment and is used by plants and planktons during photosynthesis to give out molecular atmospheric oxygen, thus completing the oxygen cycle. Therefore, suffice is to say that oxygen enters organisms in the biosphere through respiration and is expelled through photosynthesis in a process that is interconnected with the carbon cycle plus the water cycle.

However, the continued release of carbon dioxide into the atmosphere by burning fossil fuels and automobile pollution affects the oxygen cycle.

Lithosphere (Earth's Crust)

The lithosphere carries the largest quantity of all earth's oxygen, about 99.5%, because it is a constituent of the earth's lands, soils, organic matter, biomass, water, and rocks. Mostly, these constituents of the earth fix oxygen in mineral chemicals compounds such as oxides and silicates. The process is natural and happens automatically as the pure mineral elements absorb or react with the free oxygen. It happens similar to the manner in which iron picks up oxygen from the air, resulting in the formation of rust (iron oxides).

As such, during chemical reactions and some weathering processes, a portion of the trapped oxygen in the minerals is released into the atmosphere. Also, as animals and plants draw nutrient minerals from rocks, organic matter, or biomass, some of the trapped oxygen is freed in the process. Dissolved oxygen is also present in water system which is essential for the survival of aquatic life forms. As a result, these processes combined gives rise to oxygen cycle in the biosphere and lithosphere.

Processes that use Oxygen

1. Respiration: When we breather, we use oxygen and release carbon dioxide. Similarly, animals and plants also use oxygen when they breathe.

2. Combustion: When you burn something let's say a paper, you need three things for combustion to take place i.e. oxygen, fuel and heat. So, when you burn a paper, it uses oxygen and releases carbon dioxide and may be some other gases.

3. Decomposition: Decomposition occurs when plants and animals die. When this happens, they decompose and such process uses oxygen and releases carbon dioxide.

4. Rusting: When things rust, they use oxygen. This is also called as oxidation.

Processes that Produce Oxygen

1. Plants: Plants produce oxygen during the process of photosynthesis. During the process of photosynthesis, plants use carbon dioxide, sunlight, and water to create energy.

2. Sunlight: Some oxygen also gets produced when sunlight reacts with water vapor in the atmosphere.

The oxygen cycle allows for the regeneration of freely available diatomic oxygen (O_2) in the atmosphere. Oxygen accounts by volume for approximately 21 percent of the atmosphere, is reactive with myriad inorganic and organic substances, and is vital to living organisms for aerobic respiration and energy production. The cycle involves any source of oxygen within the world, and is not limited to the oxygen animals must breathe to sustain life; any compound containing an atom of oxygen is considered part of the oxygen cycle. Furthermore, the cycle is composed of many distinct biological and geological chemical reactions that together allow oxygen initially consumed and lost from the atmosphere to be released back into the atmosphere.

These reactions take place among the three different primary reservoirs, or storage areas of all of Earth's oxygen. These storage areas are varied and differ in physical and chemical form. The lithosphere, which contains the vast majority of the Earth's total oxygen, comprises the entirety of the Earth's crust and the uppermost portion of the mantle (tectonic plates can be viewed as lithospheric plates); in this reservoir, oxygen is bound in the form of rocks and minerals, primarily in silica (SiO_2) and alumina (Al_2O_3). The second reservoir is the biosphere, in which all living matter resides, including bacteria, plant life, animals, and human beings. The oxygen bound in this reservoir is found in the macromolecules of life, including nucleic acids, carbohydrates, proteins, and water. The last oxygen reservoir is the atmosphere, which is composed of approximately 20.95 percent oxygen gas, .038 percent carbon dioxide (CO_2), and water vapor (H_2O).

All of the reactions that drive the oxygen cycle occur between any two different reservoirs of oxygen. However, two chief reactions account for most of the activity in the use and regeneration of oxygen on Earth; these are cellular respiration and photosynthesis, both of which occur between the atmospheric reservoir and the biospheric reservoir. Other notable reactions contributing to the cycle include the commonplace reaction between the atmosphere's free oxygen and the lithosphere in the form of the oxidation of minerals and carbon dioxide emissions (for example, from volcanic eruptions). The biosphere and lithosphere interact in the oxygen cycle through weathering and absorbed soil nutrients for organisms and deposition of organism shells and bones into the lithosphere.

The primary action of the oxygen cycle occurs through photosynthesis and cellular respiration. Together, they represent the ability of diatomic oxygen gas to be used and produced by living organisms from the oxygen present in the air. Cellular respiration is the process by which an organism consuming food generates energy to sustain life. Most organisms use oxygen to maximize the amount of energy the food can yield, using oxygen as an electron acceptor in the electron transport chain, and using oxidative phosphorylation. Photosynthesis involves combining atmospheric carbon dioxide with water to generate carbohydrate sources and oxygen gas. The reactions are complementary:

Cellular Respiration: $C_6H_{12}O_2 + O_2 \rightarrow CO_2 + H_2O$ and

Photosynthesis: $CO_2 + H_2O \rightarrow C_6H_{12}O_2 + O_2$

There are other reactions involving oxygen that are vital for life that occur within a single reservoir. The ozone-oxygen reaction allow for stratospheric ozone to absorb ultraviolet light radiated from the sun, in addition to the visible light photosynthesis requires, which can cause damage to DNA, and break other important chemical bonds. The net reaction for the generation and breakdown of ozone by the absorption of ultraviolet light is:

- O_3 (ozone) + UV Light Energy $\rightarrow O_2$ + O (reactive radical)
- O (reactive radical) + O_2 (atmospheric oxygen gas) $\rightarrow O_3$ (ozone)

Other important reactions include the dissolving of oxygen in bodies of water, which allows for aquatic life to undergo aerobic cellular respiration and the evaporation of different bodies of water.

The oxygen cycle is not an isolated system; rather, it has intimate links with many other geological and biological cycles, such as the carbon cycle and the hydrologic cycle. Additionally, other smaller cycling systems, such as the nitrogen and sulfur cycles require oxygen from the atmosphere, living organisms, or the ground to make their vital nutrients available for uptake and binding to useful substrates. For example, the oxidation of sulfur creates sulfur dioxide, which may then react with water vapor to form sulfuric acid, which can delivered to, and taken up by, plant roots; the sulfur can be used to make cysteine, an important structural amino acid.

Nitrogen Cycle

Nitrogen is one of the primary nutrients critical for the survival of all living organisms. Although nitrogen is very abundant in the atmosphere, it is largely inaccessible in this form to most organisms.

It is largely inaccessible in this form to most organisms, making nitrogen a scarce resource and often limiting primary productivity in many ecosystems. Only when nitrogen is converted from dinitrogen gas into ammonia (NH_3) does it become available to primary producers, such as plants.

In addition to N_2 and NH_3, nitrogen exists in many different forms, including both inorganic (e.g., ammonia, nitrate) and organic (e.g., amino and nucleic acids) forms. Thus, nitrogen undergoes many different transformations in the ecosystem, changing from one form to another as organisms use it for growth and, in some cases, energy. The major transformations of nitrogen are nitrogen fixation, nitrification, denitrification, anammox, and ammonification. The transformation of nitrogen into its many oxidation states is key to productivity in the biosphere and is highly dependent on the activities of a diverse assemblage of microorganisms, such as bacteria, archaea, and fungi.

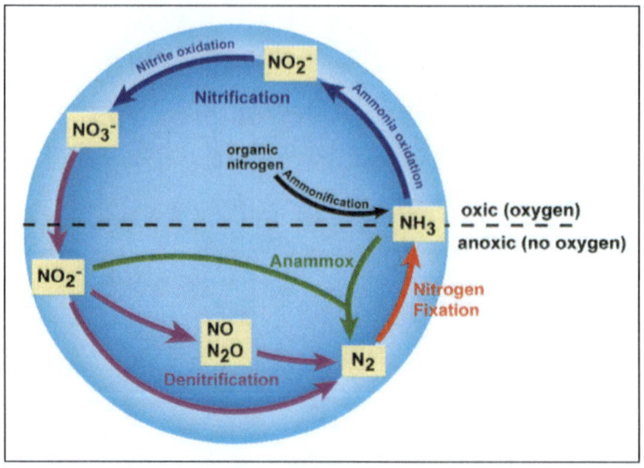

Major transformations in the nitrogen cycle.

Since the mid-1900s, humans have been exerting an ever-increasing impact on the global nitrogen cycle. Human activities, such as making fertilizers and burning fossil fuels, have significantly altered the amount of fixed nitrogen in the Earth's ecosystems. In fact, some predict that by 2030, the amount of nitrogen fixed by human activities will exceed that fixed by microbial processes. Increases in available nitrogen can alter ecosystems by increasing primary productivity and impacting carbon storage. Because of the importance of nitrogen in all ecosystems and the significant impact from human activities, nitrogen and its transformations have received a great deal of attention from ecologists.

Nitrogen Fixation

Nitrogen gas (N_2) makes up nearly 80% of the Earth's atmosphere, yet nitrogen is often the nutrient that limits primary production in many ecosystems. Why is this so? Because plants and animals are not able to use nitrogen gas in that form. For nitrogen to be available to make proteins, DNA, and other biologically important compounds, it must first be converted into a different chemical form. The process of converting N_2 into biologically available nitrogen is called nitrogen fixation. N_2 gas is a very stable compound due to the strength of the triple bond between the nitrogen atoms, and it requires a large amount of energy to break this bond. The whole process requires eight electrons and at least sixteen ATP molecules. As a result, only a select group of prokaryotes are able to carry out this energetically demanding process. Although most nitrogen fixation is carried out by prokaryotes, some nitrogen can be fixed abiotically by lightning or certain industrial processes, including the combustion of fossil fuels.

Chemical reaction of nitrogen fixation:

$$N_2 + 8\,H^+ + 8e^- \rightarrow 2NH_3 + H_2$$

Some nitrogen-fixing organisms are free-living while others are symbiotic nitrogen-fixers, which require a close association with a host to carry out the process. Most of the symbiotic associations are very specific and have complex mechanisms that help to maintain the symbiosis. For example, root exudates from legume plants (e.g., peas, clover, soybeans) serve as a signal to certain species of *Rhizobium*, which are nitrogen-fixing bacteria. This signal attracts the bacteria to the roots, and a very complex series of events then occurs to initiate uptake of the bacteria into the root and trigger the process of nitrogen fixation in nodules that form on the roots.

Some of these bacteria are aerobic, others are anaerobic; some are phototrophic, others are chemotrophic (i.e., they use chemicals as their energy source instead of light). Although there is great physiological and phylogenetic diversity among the organisms that carry out nitrogen fixation, they all have a similar enzyme complex called nitrogenase that catalyzes the reduction of N_2 to NH_3 (ammonia), which can be used as a genetic marker to identify the potential for nitrogen fixation. One of the characteristics of nitrogenase is that the enzyme complex is very sensitive to oxygen and is deactivated in its presence. This presents an interesting dilemma for aerobic nitrogen-fixers and particularly for aerobic nitrogen-fixers that are also photosynthetic since they actually produce oxygen. Over time, nitrogen-fixers have evolved different ways to protect their nitrogenase from oxygen. For example, some cyanobacteria have structures called heterocysts that provide a low-oxygen environment for the enzyme and serves as the site where all the nitrogen fixation occurs in these organisms. Other photosynthetic nitrogen-fixers fix nitrogen only at night when their photosystems are dormant and are not producing oxygen.

Genes for nitrogenase are globally distributed and have been found in many aerobic habitats (e.g., oceans, lakes, soils) and also in habitats that may be anaerobic or microaerophilic (e.g., termite guts, sediments, hypersaline lakes, microbial mats, planktonic crustaceans). The broad distribution of nitrogen-fixing genes suggests that nitrogen-fixing organisms display a very broad range of environmental conditions, as might be expected for a process that is critical to the survival of all life on Earth.

Nitrogen-fixing nodules on a clover plant root.

Table: Representative prokaryotes known to carry out nitrogen fixation.

Genus	Phylogenetic Affiliation	Lifestyle
Nostos, Anabaena	Bacteria (Cyanobacteria)	Free-living, aerobic, Phototrophic
Pseudomonas, Azotobacter, Methylomonas	Bacteria	Free-living, aerobic, Chemoorganotrophic
Alcaligenes, Thiobacillus	Bacteria	Free-living, aerobic, chemolithotrophic
Methanosarcina, Methanococcus	Archaea	Free-living, anaerobic, chemolithotrophic
Chromatium, Chlorobium	Bacteria	Free-living, anaerobic, phototrophic
Desulfovibrio, Clostridium	Bacteria	Free-living, anerobic, chemoorganotrophic
Rhizobium, Frankia	Bacteria	Symbiotic, aerobic, chemoorganotrophic

Nitrification

Nitrification is the process that converts ammonia to nitrite and then to nitrate and is another important step in the global nitrogen cycle. Most nitrification occurs aerobically and is carried out exclusively by prokaryotes. There are two distinct steps of nitrification that are carried out by distinct types of microorganisms. The first step is the oxidation of ammonia to nitrite, which is carried out by microbes known as ammonia-oxidizers. Aerobic ammonia oxidizers convert ammonia to nitrite via the intermediate hydroxylamine, a process that requires two different enzymes, ammonia monooxygenase and hydroxylamine oxidoreductase. The process generates a very small amount of energy relative to many other types of metabolism; as a result, nitrosofiers are notoriously very slow growers. Additionally, aerobic ammonia oxidizers are also autotrophs, fixing carbon dioxide to produce organic carbon, much like photosynthetic organisms, but using ammonia as the energy source instead of light.

Chemical reactions of ammonia oxidation carried out by bacteria:

$$NH_3 + O_2 + 2e^- \rightarrow NH_2OH + H_2O$$
$$NH_2OH + H_2O \rightarrow NO_2^- + 5H^+ + 4e^-$$

First reaction 1 convert's ammonia to the intermediate, hydroxylamine, and is catalyzed by the enzyme ammonia monooxygenase. Second Reaction converts hydroxylamine to nitrite and is catalyzed by the enyzmer hydroxylamine oxidoreductase.

Unlike nitrogen fixation that is carried out by many different kinds of microbes, ammonia oxidation is less broadly distributed among prokaryotes. Until recently, it was thought that all ammonia oxidation was carried out by only a few types of bacteria in the genera *Nitrosomonas*, *Nitrosospira*, and *Nitrosococcus*. However, in 2005 an archaeon was discovered that could also oxidize ammonia. Since their discovery, ammonia-oxidizing Archaea have often been found to outnumber the ammonia-oxidizing Bacteria in many habitats. In the past several years, ammonia-oxidizing Archaea have been found to be abundant in oceans, soils, and salt marshes, suggesting an important role in the nitrogen cycle for these newly-discovered organisms. Currently, only one ammonia-oxidizing archaeon has been grown in pure culture, *Nitrosopumilus maritimus*, so our understanding of their physiological diversity is limited.

The second step in nitrification is the oxidation of nitrite (NO_2^-) to nitrate (NO_3^-). This step is carried out by a completely separate group of prokaryotes, known as nitrite-oxidizing Bacteria. Some of the genera involved in nitrite oxidation include *Nitrospira*, *Nitrobacter*, *Nitrococcus*, and *Nitrospina*. Similar to ammonia oxidizers, the energy generated from the oxidation of nitrite to nitrate is very small, and thus growth yields are very low. In fact, ammonia- and nitrite-oxidizers must oxidize many molecules of ammonia or nitrite in order to fix a single molecule of CO_2. For complete nitrification, both ammonia oxidation and nitrite oxidation must occur.

Chemical reaction of nitrite oxidation:

$$NO_2^- + \frac{1}{2}O_2 \rightarrow NO_3^-$$

Ammonia-oxidizers and nitrite-oxidizers are ubiquitous in aerobic environments. They have been extensively studied in natural environments such as soils, estuaries, lakes, and open-ocean environments. However, ammonia- and nitrite-oxidizers also play a very important role in wastewater treatment facilities by removing potentially harmful levels of ammonium that could lead to the pollution of the receiving waters. Much research has focused on how to maintain stable populations of these important microbes in wastewater treatment plants. Additionally, ammonia- and nitrite-oxidizers help to maintain healthy aquaria by facilitating the removal of potentially toxic ammonium excreted in fish urine.

Anammox

Traditionally, all nitrification was thought to be carried out under aerobic conditions, but recently a new type of ammonia oxidation occurring under anoxic conditions was discovered. Anammox (anaerobic ammonia oxidation) is carried out by prokaryotes belonging to the Planctomycetes phylum of Bacteria. The first described anammox bacterium was Brocadia anammoxidans. Anammox bacteria oxidize ammonia by using nitrite as the electron acceptor to produce gaseous nitrogen. Anammox bacteria were first discovered in anoxic bioreactors of wasterwater treatment plants but have since been found in a variety of aquatic systems, including low-oxygen zones of the ocean, coastal and estuarine sediments, mangroves, and freshwater lakes. In some areas of the ocean, the anammox process is considered to be responsible for a significant loss of nitrogen. However, Ward et al. argue that denitrification rather than anammox is responsible for most nitrogen loss in other areas. Whether anammox or denitrification is responsible for most nitrogen loss in the ocean, it is clear that anammox represents an important process in the global nitrogen cycle.

Chemical reaction of anaerobic ammonia oxidation (anammox):

$$NH_4^+ + NO_2^- \rightarrow N_2 + 2H_2O$$

Denitrification

Denitrification is the process that converts nitrate to nitrogen gas, thus removing bioavailable nitrogen and returning it to the atmosphere. Dinitrogen gas (N_2) is the ultimate end product of denitrification, but other intermediate gaseous forms of nitrogen exist. Some of these gases, such as nitrous oxide (N_2O), are considered greenhouse gasses, reacting with ozone and contributing to air pollution.

Reactions involved in denitrification: First reaction represents the steps of reducing nitrate to dinitrogen gas. Second reaction represents the complete redox reaction of denitrification:

$$NO_3^- \rightarrow NO_2^- \rightarrow NO + N_2O \rightarrow N_2$$

$$2\ NO_3^- + 10e^- + 12H^+ \rightarrow N_2 + 6H_2O$$

Unlike nitrification, denitrification is an anaerobic process, occurring mostly in soils and sediments and anoxic zones in lakes and oceans. Similar to nitrogen fixation, denitrification is carried out by a diverse group of prokaryotes, and there is recent evidence that some eukaryotes are also capable of denitrification. Some denitrifying bacteria include species in the genera *Bacillus*, *Paracoccus*, and *Pseudomonas*. Denitrifiers are chemoorganotrophs and thus must also be supplied with some form of organic carbon.

Denitrification is important in that it removes fixed nitrogen (i.e., nitrate) from the ecosystem and returns it to the atmosphere in a biologically inert form (N_2). This is particularly important in agriculture where the loss of nitrates in fertilizer is detrimental and costly. However, denitrification in wastewater treatment plays a very beneficial role by removing unwanted nitrates from the wastewater effluent, thereby reducing the chances that the water discharged from the treatment plants will cause undesirable consequences (e.g., algal blooms).

Ammonification

When an organism excretes waste or dies, the nitrogen in its tissues is in the form of organic nitrogen (e.g. amino acids, DNA). Various fungi and prokaryotes then decompose the tissue and release inorganic nitrogen back into the ecosystem as ammonia in the process known as ammonification. The ammonia then becomes available for uptake by plants and other microorganisms for growth.

Ecological Implications of Human Alterations to the Nitrogen Cycle

Many human activities have a significant impact on the nitrogen cycle. Burning fossil fuels, application of nitrogen-based fertilizers, and other activities can dramatically increase the amount of biologically available nitrogen in an ecosystem. And because nitrogen availability often limits the primary productivity of many ecosystems, large changes in the availability of nitrogen can lead to severe alterations of the nitrogen cycle in both aquatic and terrestrial ecosystems. Industrial nitrogen fixation has increased exponentially since the 1940s, and human activity has doubled the amount of global nitrogen fixation.

In terrestrial ecosystems, the addition of nitrogen can lead to nutrient imbalance in trees, changes in forest health, and declines in biodiversity. With increased nitrogen availability there is often a change in carbon storage, thus impacting more processes than just the nitrogen cycle. In agricultural systems, fertilizers are used extensively to increase plant production, but unused nitrogen, usually in the form of nitrate, can leach out of the soil, enter streams and rivers, and ultimately make its way into our drinking water. The process of making synthetic fertilizers for use in agriculture by causing N_2 to react with H_2, known as the Haber-Bosch process, has increased significantly over the past several decades. In fact, today, nearly 80% of the nitrogen found in human tissues originated from the Haber-Bosch process.

Much of the nitrogen applied to agricultural and urban areas ultimately enters rivers and nearshore coastal systems. In nearshore marine systems, increases in nitrogen can often lead to anoxia (no oxygen) or hypoxia (low oxygen), altered biodiversity, changes in food-web structure, and general habitat degradation. One common consequence of increased nitrogen is an increase in harmful algal blooms. Toxic blooms of certain types of dinoflagellates have been associated with high fish and shellfish mortality in some areas. Even without such economically catastrophic effects, the addition of nitrogen can lead to changes in biodiversity and species composition that may lead to changes in overall ecosystem function. Some have even suggested that alterations to the nitrogen cycle may lead to an increased risk of parasitic and infectious diseases among humans and wildlife. Additionally, increases in nitrogen in aquatic systems can lead to increased acidification in freshwater ecosystems.

Marine Nitrogen Cycle

The nitrogen cycle is an important process in the ocean as well. While the overall cycle is similar, there are different players and modes of transfer for nitrogen in the ocean. Nitrogen enters the water through the precipitation, runoff, or as N_2 from the atmosphere. Nitrogen cannot be utilized by phytoplankton as N_2 so it must undergo nitrogen fixation which is performed predominately by cyanobacteria. Without supplies of fixed nitrogen entering the marine cycle, the fixed nitrogen would be used up in about 2000 years. Phytoplankton needs nitrogen in biologically available forms for the initial synthesis of organic matter. Ammonia and urea are released into the water by excretion from plankton. Nitrogen sources are removed from the euphotic zone by the downward movement of the organic matter. This can occur from sinking of phytoplankton, vertical mixing, or sinking of waste of vertical migrators. The sinking results in ammonia being introduced at lower depths below the euphotic zone. Bacteria are able to convert ammonia to nitrite and nitrate but they are inhibited by light so this must occur below the euphotic zone. Ammonification or Mineralization is performed by bacteria to convert organic nitrogen to ammonia. Nitrification can then occur to convert the ammonium to nitrite and nitrate. Nitrate can be returned to the euphotic zone by vertical mixing and upwelling where it can be taken up by phytoplankton to continue the cycle. N_2 can be returned to the atmosphere through denitrification.

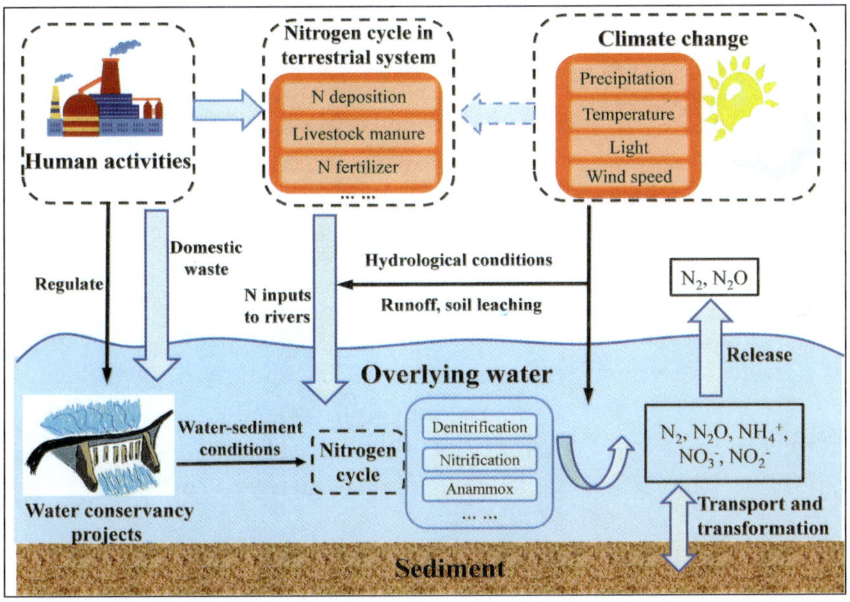

A schematic representing the Marine Nitrogen Cycle.

Ammonium is thought to be the preferred source of fixed nitrogen for phytoplankton because its assimilation does not involve a redox reaction and therefore requires little energy. Nitrate requires a redox reaction for assimilation but is more abundant so most phytoplanktons have adapted to have the enzymes necessary to undertake this reduction (nitrate reductase). There are a few notable and well-known exceptions that include *Prochlorococcus* and some *Synechococcus*. These species can only take up nitrogen as ammonium.

The nutrients in the ocean are not uniformly distributed. Areas of upwelling provide supplies of nitrogen from below the euphotic zone. Coastal zones provide nitrogen from runoff and upwelling occurs readily along the coast. However, the rate at which nitrogen can be taken up by phytoplankton is decreased in oligotrophic waters year-round and temperate water in the summer resulting in lower primary production. The distribution of the different forms of nitrogen varies throughout the oceans as well.

Nitrate is depleted in near-surface water except in upwelling regions. Coastal upwelling regions usually have high nitrate and chlorophyll levels as a result of the increased production. However, there are regions of high surface nitrate but low chlorophyll that are referred to as HNLC (high nitrogen, low chlorophyll) regions. The best explanation for HNLC regions relates to iron scarcity in the ocean, which may play an important part in ocean dynamics and nutrient cycles. The input of iron varies by region and is delivered to the ocean by dust (from dust storms) and leached out of rocks. Iron is under consideration as the true limiting element to ecosystem productivity in the ocean.

Ammonium and nitrite show a maximum concentration at 50–80 m (lower end of the euphotic zone) with decreasing concentration below that depth. This distribution can be accounted for by the fact that nitrite and ammonium are intermediate species. They are both rapidly produced and consumed through the water column. The amount of ammonium in the ocean is about 3 orders of magnitude less than nitrate. Between ammonium, nitrite, and nitrate, nitrite has the fastest turnover rate. It can be produced during nitrate assimilation, nitrification, and denitrification; however, it is immediately consumed again.

New vs. Regenerated Nitrogen

Nitrogen entering the euphotic zone is referred to as new nitrogen because it is newly arrived from outside the productive layer. The new nitrogen can come from below the euphotic zone or from outside sources. Outside sources are upwelling from deep water and nitrogen fixation. If the organic matter is eaten, respired, delivered to the water as ammonia, and re-incorporated into organic matter by phytoplankton it is considered recycled/regenerated production.

New production is an important component of the marine environment. One reason is that only continual input of new nitrogen can determine the total capacity of the ocean to produce a sustainable fish harvest. Harvesting fish from regenerated nitrogen areas will lead to a decrease in nitrogen and therefore a decrease in primary production. This will have a negative effect on the system. However, if fish are harvested from areas of new nitrogen the nitrogen will be replenished.

Human Influences on the Nitrogen Cycle

As a result of extensive cultivation of legumes (particularly soy, alfalfa, and clover), growing use of the Haber–Bosch process in the creation of chemical fertilizers, and pollution emitted by vehicles and industrial plants, human beings have more than doubled the annual transfer of nitrogen into biologically available forms. In addition, humans have significantly contributed to the transfer of nitrogen trace gases from Earth to the atmosphere and from the land to aquatic systems. Human alterations to the global nitrogen cycle are most intense in developed countries and in Asia, where vehicle emissions and industrial agriculture are highest.

Generation of Nr, reactive nitrogen, has increased over 10 fold in the past century due to global industrialisation. This form of nitrogen follows a cascade through the biosphere via a variety of mechanisms, and is accumulating as the rate of its generation is greater than the rate of denitrification.

Nitrous oxide (N_2O) has risen in the atmosphere as a result of agricultural fertilization, biomass burning, cattle and feedlots, and industrial sources. N_2O has deleterious effects in the stratosphere, where it breaks down and acts as a catalyst in the destruction of atmospheric ozone. Nitrous oxide is also a greenhouse gas and is currently the third largest contributor to global warming, after carbon dioxide and methane. While not as abundant in the atmosphere as carbon dioxide, it is, for an equivalent mass, nearly 300 times more potent in its ability to warm the planet.

Ammonia (NH_3) in the atmosphere has tripled as the result of human activities. It is a reactant in the atmosphere, where it acts as an aerosol, decreasing air quality and clinging to water droplets, eventually resulting in nitric acid (HNO_3) that produces acid rain. Atmospheric ammonia and nitric acid also damage respiratory systems.

The very high temperature of lightning naturally produces small amounts of NO_x, NH_3, and HNO_3, but high-temperature combustion has contributed to a 6- or 7-fold increase in the flux of NO_x to the atmosphere. Its production is a function of combustion temperature - the higher the temperature, the more NO_x is produced. Fossil fuel combustion is a primary contributor, but so are biofuels and even the burning of hydrogen. However, the rate that hydrogen is directly injected into the combustion chambers of internal combustion engines can be controlled to prevent the higher combustion temperatures that produce NO_x.

Ammonia and nitrous oxides actively alter atmospheric chemistry. They are precursors of tropospheric (lower atmosphere) ozone production, which contributes to smog and acid rain, damages plants and increases nitrogen inputs to ecosystems. Ecosystem processes can increase with nitrogen fertilization, but anthropogenic input can also result in nitrogen saturation, which weakens productivity and can damage the health of plants, animals, fish, and humans.

Decreases in biodiversity can also result if higher nitrogen availability increases nitrogen-demanding grasses, causing a degradation of nitrogen-poor, species-diverse heathlands.

Consequence of Human Modification of the Nitrogen Cycle

Impacts on Natural Systems

Increasing levels of nitrogen deposition are shown to have a number of negative effects on both terrestrial and aquatic ecosystems. Nitrogen gases and aerosols can be directly toxic to certain plant species, affecting the aboveground physiology and growth of plants near large point sources of nitrogen pollution. Changes to plant species may also occur, as accumulation of nitrogen compounds increase its availability in a given ecosystem, eventually changing the species composition, plant diversity, and nitrogen cycling. Ammonia and ammonium - two reduced forms of nitrogen - can be detrimental over time due to an increased toxicity toward sensitive species of plants, particularly those that are accustomed to using nitrate as their source of nitrogen, causing poor development of their roots and shoots. Increased nitrogen deposition also leads to soil acidification, which increases base cation leaching in the soil and amounts of aluminum and other potentially toxic metals, along with decreasing the amount of nitrification occurring and increasing plant-derived litter. Due to the ongoing changes caused by high nitrogen deposition, an environment's susceptibility to ecological stress and disturbance - such as pests and pathogens - may increase, thus making it less resilient to situations that otherwise would have little impact to its long-term vitality.

Additional risks posed by increased availability of inorganic nitrogen in aquatic ecosystems include water acidification; eutrophication of fresh and saltwater systems; and toxicity issues for animals, including humans. Eutrophication often leads to lower dissolved oxygen levels in the water column, including hypoxic and anoxic conditions, which can cause death of aquatic fauna. Relatively sessile benthoses, or bottom-dwelling creatures, are particularly vulnerable because of their lack of mobility, though large fish kills are not uncommon. Oceanic dead zones near the mouth of the Mississippi in the Gulf of Mexico are a well-known example of algal bloom-induced hypoxia. The New York Adirondack Lakes, Catskills, Hudson Highlands, Rensselaer Plateau and parts of Long Island display the impact of nitric acid rain deposition, resulting in the killing of fish and many other aquatic species.

Ammonia (NH_3) is highly toxic to fish and the level of ammonia discharged from wastewater treatment facilities must be closely monitored. To prevent fish deaths, nitrification via aeration prior to discharge is often desirable. Land application can be an attractive alternative to the aeration.

Impacts on Human Health: Nitrate Accumulation in Drinking Water

Leakage of Nr (reactive nitrogen) from human activities can cause nitrate accumulation in the natural water environment, which can create harmful impacts on human health. Excessive use of N-fertilizer in agriculture has been one of the major sources of nitrate pollution in groundwater and surface water. Due to its high solubility and low retention by soil, nitrate can easily escape from the subsoil layer to the groundwater, causing nitrate pollution. Some other non-point sources for nitrate pollution in groundwater are originated from livestock feeding, animal and human contamination and municipal and industrial waste. Since groundwater often serves as the primary domestic water supply, nitrate pollution can be extended from groundwater to surface and drinking water in the process of potable water production, especially for small community water supplies, where poorly regulated and unsanitary waters are used.

The WHO standard for drinking water is 50 mg NO_3^- L^{-1} for short-term exposure, and for 3 mg NO_3^- L^{-1}chronic effects. Once it enters human body, nitrate can react with organic compounds through nitrosation reactions in the stomach to form nitrosamines and nitrosamides, which are involved in some types of cancers (e.g., oral cancer and gastric cancer).

Impacts on Human Health: Air Quality

Human activities have also dramatically altered the global nitrogen cycle via production of nitrogenous gases, associated with the global atmospheric nitrogen pollution. There are multiple sources of atmospheric reactive nitrogen (Nr) fluxes. Agricultural sources of reactive nitrogen can produce atmospheric emission of ammonia ($NH_{3)}$, nitrogen oxides (NO_x) and nitrous oxide (N_2O). Combustion processes in energy production, transportation and industry can also result in the formation of new reactive nitrogen via the emission of NO_x, an unintentional waste product. When those reactive nitrogens are released to the lower atmosphere, they can induce the formation of smog, particulate matter and aerosols, all of which are major contributors to adverse health effects on human health from air pollution. In the atmosphere, NO_2 can be oxidized to nitric acid (HNO_3), and it can further react with NH_3 to form ammonium nitrate, which facilitates the formation of particular nitrate. Moreover, NH_3 can react with other acid gases (sulfuric and hydrochloric acids) to form ammonium-containing particles, which are the precursors for the secondary organic aerosol particles in photochemical smog.

Sulfur Cycle

Sulfur (S), the tenth most abundant element in the universe, is a brittle, yellow, tasteless, and odorless non-metallic element. It comprises many vitamins, proteins, and hormones that play critical roles in both climate and in the health of various ecosystems. The majority of the Earth's sulfur is stored underground in rocks and minerals, including as sulfate salts buried deep within ocean sediments.

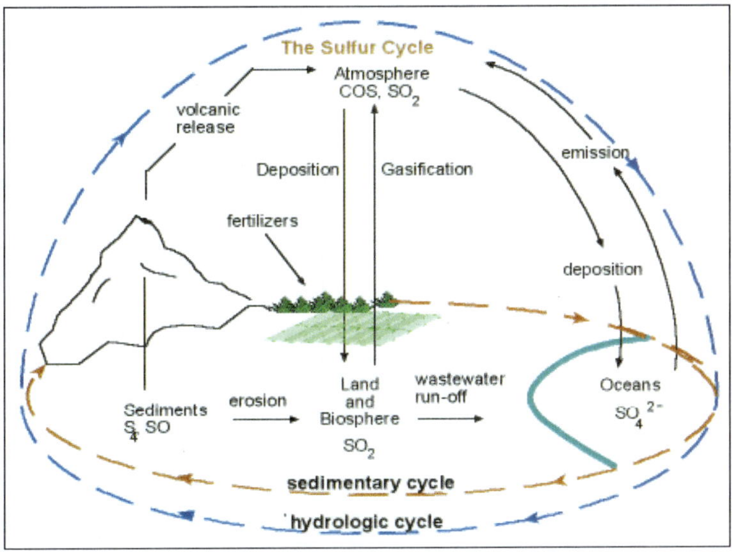

The sulfur cycle contains both atmospheric and terrestrial processes. Within the terrestrial portion, the cycle begins with the weathering of rocks, releasing the stored sulfur. The sulfur then comes into contact with air where it is converted into sulfate (SO_4). The sulfate is taken up by plants and microorganisms and is converted into organic forms; animals then consume these organic forms through foods they eat, thereby moving the sulfur through the food chain. As organisms die and decompose, some of the sulfur is again released as a sulfate and some enters the tissues of microorganisms. There are also a variety of natural sources that emit sulfur directly into the atmosphere, including volcanic eruptions, the breakdown of organic matter in swamps and tidal flats, and the evaporation of water.

Sulfur eventually settles back into the Earth or comes down within rainfall. A continuous loss of sulfur from terrestrial ecosystem runoff occurs through drainage into lakes and streams, and eventually oceans. Sulfur also enters the ocean through fallout from the Earth's atmosphere. Within the ocean, some sulfur cycles through marine communities, moving through the food chain. A portion of this sulfur is emitted back into the atmosphere from sea spray. The remaining sulfur is lost to the ocean depths, combining with iron to form ferrous sulfide which is responsible for the black color of most marine sediments.

Since the Industrial Revolution, human activities have contributed to the amount of sulfur that enters the atmosphere, primarily through the burning of fossil fuels and the processing of metals. One-third of all sulfur that reaches the atmosphere—including 90% of sulfur dioxide—stems from human activities. Emissions from these activities, along with nitrogen emissions, react with other chemicals in the atmosphere to produce tiny particles of sulfate salts which fall as acid rain, causing a variety of damage to both the natural environment as well as to man-made environments, such as the chemical weathering of buildings. However, as particles and tiny airborne droplets, sulfur also acts as a regulator of global climate. Sulfur dioxide and sulfate aerosols absorb ultraviolet radiation, creating cloud cover that cools cities and may offset global warming caused by the greenhouse effect.

Sulfur Sources and Sinks

Sulfur is found in oxidation states ranging from +6 in SO_4^{2-} to -2 in sulfides. Thus, elemental sulfur can either give or receive electrons depending on its environment. On the anoxic early Earth, most sulfur was present in minerals such as pyrite (FeS_2). Over Earth history, the amount of mobile sulfur increased through volcanic activity as well as weathering of the crust in an oxygenated atmosphere. Earth's main sulfur sink is the oceans SO_4^{2-}, where it is the major oxidizing agent.

When SO_4^{2-} is assimilated by organisms, it is reduced and converted to organic sulfur, which is an essential component of proteins. However, the biosphere does not act as a major sink for sulfur, instead the majority of sulfur is found in seawater or sedimentary rocks including: pyrite rich shales, evaporite rocks (anhydrite and baryte), and calcium and magnesium carbonates (i.e. carbonate-associated sulfate). The amount of sulfate in the oceans is controlled by three major processes:

1. Input from rivers.
2. Sulfate reduction and sulfide re-oxidation on continental shelves and slopes.
3. Burial of anhydrite and pyrite in the oceanic crust.

The primary natural source of sulfur to the atmosphere is sea spray or windblown sulfur rich dust, neither of which is long lived in the atmosphere. In recent times, the large annual input of sulfur from the burning of coal and other fossil fuels has added a substantial amount SO_2 which acts as an air pollutant. In the geologic past, igneous intrusions into coal measures have caused large scale burning of these measures, and consequential release of sulfur to the atmosphere. This has led to substantial disruption to the climate system, and is one of the proposed causes of the Permian–Triassic extinction event.

Dimethylsulfide [$(CH_3)_2S$ or DMS] is produced by the decomposition of dimethylsulfoniopropionate (DMSP) from dying phytoplankton cells in the ocean's photic zone, and is the major biogenic gas emitted from the sea, where it is responsible for the distinctive "smell of the sea" along coastlines. DMS is the largest natural source of sulfur gas, but still only has a residence time of about one day in the atmosphere and a majority of it is redeposited in the oceans rather than making it to land. However, it is a significant factor in the climate system, as it is involved in the formation of clouds.

Biologically and Thermochemically Driven Sulfate Reduction

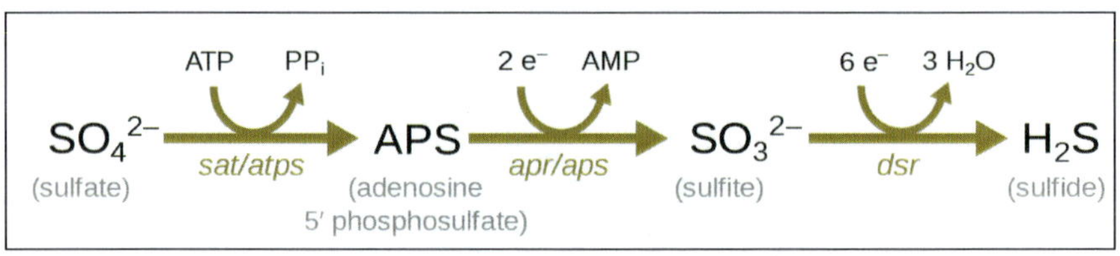

Dissimilatory sulfate reduction pathway.

Sulfur can be reduced both biologically and thermochemically. Dissimilatory sulfate reduction has two different definitions:

1. The microbial process that converts sulfate to sulfide for energy gain, and
2. A set of forward and reverse pathways that progress from the uptake and release of sulfate by the cell to its conversion to various sulfur intermediates, and ultimately to sulfide which is released from the cell.

Sulfide and thiosulfate are the most abundant reduced inorganic sulfur species in the environments and are converted to sulfate, primarily by bacterial action, in the oxidative half of the sulfur cycle. Bacterial sulfate reduction (BSR) usually occurs at temperatures from 0 up to 60–80 °C because above that temperature almost all sulfate-reducing microbes can no longer metabolize. However, in hot sediments around hydrothermal vents BSR can happen at temperatures up to 110 °C, due to hyperthermophilic bacteria that were discovered around these vents in the Guaymas Basin. Bacterial sulfate reduction is geologically instantaneous happening on the order of hundreds to thousands of years. Thermochemical sulfate reduction (TSR) occurs at much higher temperatures (160–180 °C) and over longer time intervals, several tens of thousands to a few million years.

The main difference between these two reactions is that they occur at different thermal regimes. Additionally, one reaction is reduced bacterially, while the other is reduced inorganically. Therefore, the temperature for thermochemical sulfate reduction is much higher due to the activation energy required to reduce sulfate. Bacterial sulfate reductions requires lower temperatures because the sulfur reducing bacteria can only live at relatively low temperature (below 60 °C). Bacterial sulfate reduction also requires a relatively open system; otherwise the bacteria will poison themselves when the sulfate levels rise above 5–10%.

The organic reactants involved in bacterial sulfate reduction are organic acids which are distinctive from the organic reactants needed for thermochemical sulfate reduction. In both cases sulfate is usually derived from the dissolution of gypsum or taken directly out of the seawater. The factors that control whether bacterial sulfate reduction or thermochemical sulfate reduction will occur are temperature, which is generally a product of depth, with bacterial sulfate reduction occurring in shallower levels than thermochemical sulfate reduction. Their solid products are similar but can be distinguished from one another petrographically, due to their differing crystal sizes, shapes and reflectivity.

Sulfur-oxidizing Bacteria in Hydrothermal Vents

Hydrothermal vents emit hydrogen sulfide that support the carbon fixation of chemolithotrophic bacteria that oxidize hydrogen sulfide with oxygen to produce elemental sulfur or sulfate. The chemical reactions are as follows:

$$CO_2 + 4H_2S + O_2 \text{ -> } CH_2O + 4S^o + 3H_2O$$

$$CO_2 + H_2S + O_2 + H_2O \text{ -> } CH_2O + SO_4^{2-} + 2H^+$$

In modern oceans, *Thiomicrospira*, *Halothiobacillus*, and *Beggiatoa* are primary sulfur oxidizing bacteria, and form chemosynthetic symbioses with animal hosts. The host provides metabolic substrates (e.g., CO_2, O_2, H_2O) to the symbiont while the symbiont generates organic carbon for sustaining the metabolic activities of the host. The produced sulfate usually combines with the leached calcium ions to form gypsum, which can form widespread deposits on near mid-ocean spreading centers.

$\delta^{34}S$

Although 25 isotopes are known for sulfur, only four are stable and of geochemical importance. Of those four, two (^{32}S, light and ^{34}S, heavy) comprise (99.22%) of S on Earth. The vast majority (95.02%) of S occurs as ^{32}S with only 4.21% in ^{34}S. The ratio of these two isotopes is fixed in our solar system and has been since its formation. The bulk Earth sulfur isotopic ratio is thought to be the same as the ratio of 22.22 measured from the Canyon Diablo troilite (CDT), a meteorite. That ratio is accepted as the international standard and is therefore set at δ0.00. Deviation from 0.00 is expressed as the $\delta^{34}S$ which is a ratio in per mill (‰). Positive values correlate to increased levels of ^{34}S, whereas negative values correlate with greater ^{32}S in a sample.

Formation of sulfur minerals through non-biogenic processes does not substantially differentiate between the light and heavy isotopes, therefore sulfur isotope ratios in gypsum or barite should be the same as the overall isotope ratio in the water column at their time of precipitation. Sulfate reduction through biologic activity strongly differentiates between the two isotopes because of the more rapid enzymic reaction with ^{32}S. Sulfate metabolism results in an isotopic depletion of -18%, and repeated cycles of oxidation and reduction can result in values up to -50 %. Average present day seawater values of $\delta^{34}S$ are on the order of +21%.

Throughout geologic history the sulfur cycle and the isotopic ratios have coevolved with the biosphere becoming overall more negative with the increases in biologically driven sulfate reduction, but also show substantial positive excursion. In general positive excursions in the sulfur isotopes mean that there is an excess of pyrite deposition rather than oxidation of sulfide minerals exposed on land.

Marine Sulfur Cycle

The sulfur cycle in marine environments has been well-studied via the tool of sulfur isotope systematics expressed as $\delta^{34}S$. The modern global oceans have sulfur storage of 1.3×10^{21} g, mainly occurring as sulfate with the $\delta^{34}S$ value of +21‰. The overall input flux is 1.0×10^{14} g/year with the sulfur isotope composition of ~3‰. Riverine sulfate derived from the terrestrial weathering of sulfide minerals ($\delta^{34}S$ = +6‰) is the primary input of sulfur to the oceans. Other sources are metamorphic and volcanic degassing and hydrothermal activity ($\delta^{34}S$ = 0‰), which release reduced sulfur species (e.g., H_2S and S^0). There are two major outputs of sulfur from the oceans. The first sink is the burial of sulfate either as marine evaporites (e.g., gypsum) or carbonate-associated sulfate (CAS), which accounts for 6×10^{13} g/year ($\delta^{34}S$ = +21‰). The second sulfur sink is pyrite burial in shelf sediments or deep seafloor sediments (4×10^{13} g/year; $\delta^{34}S$ = -20‰).The total marine sulfur output flux is 1.0×10^{14} g/year which matches the input fluxes, implying the modern marine sulfur budget is at steady state. The residence time of sulfur in modern global oceans is 13,000,000 years.

Evolution of the Sulfur Cycle

The isotopic composition of sedimentary sulfides provides primary information on the evolution of the sulfur cycle.

The total inventory of sulfur compounds on the surface of the Earth (nearly 10^{22} g S) represents the total outgassing of sulfur through geologic time. Rocks analyzed for sulfur content are generally organic-rich shales meaning they are likely controlled by biogenic sulfur reduction. Average seawater curves are generated from evaporites deposited throughout geologic time because again, since they do not discriminate between the heavy and light sulfur isotopes, they should mimic the ocean composition at the time of deposition.

4.6 billion years ago (Ga) the Earth formed and had a theoretical $\delta^{34}S$ value of 0. Since there was no biologic activity on early Earth there would be no isotopic fractionation. All sulfur in the atmosphere would be released during volcanic eruptions. When the oceans condensed on Earth, the atmosphere was essentially swept clean of sulfur gases, owing to their high solubility in water. Throughout the majority of the Archean (4.6–2.5 Ga) most systems appeared to be sulfate-limited. Some small Archean evaporite deposits require that at least locally elevated concentrations (possibly due to local volcanic activity) of sulfate existed in order for them to be supersaturated and precipitate out of solution.

3.8–3.6 Ga marks the beginning of the exposed geologic record because this is the age of the oldest rocks on Earth. Metasedimentary rocks from this time still have an isotopic value of 0 because the biosphere was not developed enough (possibly at all) to fractionate sulfur.

3.5 Ga anoxyogenic photosynthesis is established and provides a weak source of sulfate to the global ocean with sulfate concentrations incredibly low the $\delta^{34}S$ is still basically 0. Shortly after, at 3.4 Ga the first evidence for minimal fractionation in evaporitic sulfate in association with magmatically derived sulfides can be seen in the rock record. This fractionation shows possible evidence for anoxygenic phototrophic bacteria.

2.8 Ga marks the first evidence for oxygen production through photosynthesis. This is important because there cannot be sulfur oxidation without oxygen in the atmosphere. This exemplifies the coevolution of the oxygen and sulfur cycles as well as the biosphere.

2.7–2.5 Ga is the age of the oldest sedimentary rocks to have a depleted $\delta\ ^{34}S$ which provide the first compelling evidence for sulfate reduction.

2.3 Ga sulfate increases to more than 1 mM; this increase in sulfate is coincident with the "Great Oxygenation Event", when redox conditions on Earth's surface are thought by most workers to have shifted fundamentally from reducing to oxidizing. This shift would have led to an incredible increase in sulfate weathering which would have led to an increase in sulfate in the oceans. The large isotopic fractionations that would likely be associated with bacteria reduction are produced for the first time. Although there was a distinct rise in seawater sulfate at this time it was likely still only less than 5–15% of present-day levels.

At 1.8 Ga, Banded iron formations (BIF) are common sedimentary rocks throughout the Archean and Paleoproterozoic; their disappearance marks a distinct shift in the chemistry of ocean water. BIFs have alternating layers of iron oxides and chert. BIFs only form if the water is allowed to supersaturate in dissolved iron (Fe^{2+}) meaning there cannot be free oxygen or sulfur in the water column because it would form Fe^{3+} (rust) or pyrite and precipitate out of solution. Following this supersaturation, the water must become oxygenated in order for the ferric rich bands to precipitate it must still be sulfur poor otherwise pyrite would form instead of Fe^{3+}. It has been hypothesized that BIFs formed during the initial evolution of photosynthetic organisms that had phases of population growth, causing over production of oxygen. Due to this over production they would poison themselves causing a mass die off, which would cut off the source of oxygen and produce a large amount of CO_2 through the decomposition of their bodies, allowing for another bacterial bloom. After 1.8 Ga sulfate concentrations were sufficient to increase rates of sulfate reduction to greater than the delivery flux of iron to the oceans.

Along with the disappearance of BIF, the end of the Paleoproterozoic also marks the first large scale sedimentary exhalative deposits showing a link between mineralization and a likely increase in the amount of sulfate in sea water. In the Paleoproterozoic the sulfate in seawater had increased to an amount greater than in the Archean, but was still lower than present day values. The sulfate levels in the Proterozoic also act as proxies for atmospheric oxygen because sulfate is produced mostly through weathering of the continents in the presence of oxygen. The low levels in the Proterozoic simply imply that levels of atmospheric oxygen fell between the abundances of the Phanerozoic and the deficiencies of the Archean.

750 million years ago (Ma) there is a renewed deposition of BIF which marks a significant change in ocean chemistry. This was likely due to snowball earth episodes where the entire globe including the oceans was covered in a layer of ice cutting off oxygenation. In the late Neoproterozoic high carbon burial rates increased the atmospheric oxygen level to >10% of its present-day value. In the Latest Neoproterozoic another major oxidizing event occurred on Earth's surface that resulted in an oxic deep ocean and possibly allowed for the appearance of multicellular life.

During the last 600 million years, seawater SO_4 has varied between +10 and +30‰ in $\delta^{34}S$, with an average value close to that of today. This coincides with atmospheric O_2 levels reaching something close to modern values around the Precambrian–Cambrian boundary.

Over a shorter time scale (ten million years) changes in the sulfur cycle are easier to observe and can be even better constrained with oxygen isotopes. Oxygen is continually incorporated into the sulfur cycle through sulfate oxidation and then released when that sulfate is reduced once again. Since different sulfate sources within the ocean have distinct oxygen isotopic values it may be possible to use oxygen to trace the sulfur cycle. Biological sulfate reduction preferentially selects lighter oxygen isotopes for the same reason that lighter sulfur isotopes are preferred. By studying oxygen isotopes in ocean sediments over the last 10 million years were able to better constrain the sulfur concentrations in sea water through that same time. They found that the sea level changes due to Pliocene and Pleistocene glacial cycles changed the area of continental shelves which then disrupted the sulfur processing, lowering the concentration of sulfate in the sea water. This was a drastic change as compared to preglacial times before 2 million years ago.

Great Oxidation Event and Sulfur Isotope Mass-independent Fractionation

The Great Oxygenation Event (GOE) is characterized by the disappearance of sulfur isotope mass-independent fractionation (MIF) in the sedimentary records at around 2.45 billion years ago

(Ga). The MIF of sulfur isotope ($\Delta^{33}S$) is defined by the deviation of measured $\delta^{33}S$ value from the $\delta^{33}S$ value inferred from the measured $\delta^{34}S$ value according to the mass dependent fractionation law. The Great Oxidation Event represented a massive transition of global sulfur cycles. Before the Great Oxidation Event, the sulfur cycle was heavily influenced by the ultraviolet (UV) radiation and the associated photochemical reactions, which induced the sulfur isotope mass-independent fractionation ($\Delta^{33}S \neq 0$). The preservation of sulfur isotope mass-independent fractionation signals requires the atmospheric O_2 lower than 10^{-5} of present atmospheric level (PAL). The disappearance of sulfur isotope mass-independent fractionation at ~2.45 Ga indicates that atmospheric pO_2 exceeded 10^{-5} present atmospheric level after the Great Oxygenation Event. Oxygen played an essential role in the global sulfur cycles after the Great Oxygenation Event, such as oxidative weathering of sulfides. The burial of pyrite in sediments in turn contributes to the accumulation of free O_2 in Earth's surface environment.

Economic Importance

Sulfur is intimately involved in production of fossil fuels and a majority of metal deposits because of its ability to act as an oxidizing or reducing agent. The vast majority of the major mineral deposits on Earth contain a substantial amount of sulfur including, but not limited to: sedimentary exhalative deposits (SEDEX), Carbonate-hosted lead-zinc ore deposits (Mississippi Valley-Type MVT) and porphyry copper deposits. Iron sulfides, galena and sphalerite will form as by-products of hydrogen sulfide generation, as long as the respective transition or base metals are present or transported to a sulfate reduction site. If the system runs out of reactive hydrocarbons economically viable elemental sulfur deposits may form. Sulfur also acts as a reducing agent in many natural gas reservoirs and generally ore forming fluids have a close relationship with ancient hydrocarbon seeps or vents.

Important sources of sulfur in ore deposits are generally deep-seated, but they can also come from local country rocks, sea water, or marine evaporites. The presence or absence of sulfur is one of the limiting factors on both the concentration of precious metals and its precipitation from solution. pH, temperature and especially redox states determine whether sulfides will precipitate. Most sulfide brines will remain in concentration until they reach reducing conditions, a higher pH or lower temperatures.

Ore fluids are generally linked to metal rich waters that have been heated within a sedimentary basin under the elevated thermal conditions typically in extensional tectonic settings. The redox conditions of the basin lithologies exert an important control on the redox state of the metal-transporting fluids and deposits can form from both oxidizing and reducing fluids. Metal-rich ore fluids tend to be by necessity comparatively sulfide deficient, so a substantial portion of the sulfide must be supplied from another source at the site of mineralization. Bacterial reduction of seawater sulfate or a euxinic (anoxic and H_2S-containing) water column is a necessary source of that sulfide. When present, the $\delta^{34}S$ values of barite are generally consistent with a seawater sulfate source, suggesting baryte formation by reaction between hydrothermal barium and sulfate in ambient seawater.

Once fossil fuels or precious metals are discovered and either burned or milled, the sulfur become a waste product which must be dealt with properly or it can become a pollutant. There has been a great increase in the amount of sulfur in our present day atmosphere because of the burning of fossil fuels. Sulfur acts as a pollutant and an economic resource at the same time.

Human Impact

Human activities have a major effect on the global sulfur cycle. The burning of coal, natural gas, and other fossil fuels has greatly increased the amount of S in the atmosphere and ocean and depleted the sedimentary rock sink. Without human impact sulfur would stay tied up in rocks for millions of years until it was uplifted through tectonic events and then released through erosion and weathering processes. Instead it is being drilled, pumped and burned at a steadily increasing rate. Over the most polluted areas there has been a 30-fold increase in sulfate deposition.

Although the sulfur curve shows shifts between net sulfur oxidation and net sulfur reduction in the geologic past, the magnitude of the current human impact is probably unprecedented in the geologic record. Human activities greatly increase the flux of sulfur to the atmosphere, some of which is transported globally. Humans are mining coal and extracting petroleum from the Earth's crust at a rate that mobilizes 150 x 10^{12} gS/yr, which is more than double the rate of 100 years ago. The result of human impact on these processes is to increase the pool of oxidized sulfur (SO_4) in the global cycle, at the expense of the storage of reduced sulfur in the Earth's crust. Therefore, human activities do not cause a major change in the global pools of S, but they do produce massive changes in the annual flux of S through the atmosphere.

When SO_2 is emitted as an air pollutant, it forms sulfuric acid through reactions with water in the atmosphere. Once the acid is completely dissociated in water the pH can drop to 4.3 or lower causing damage to both man-made and natural systems. According to the EPA, acid rain is a broad term referring to a mixture of wet and dry deposition (deposited material) from the atmosphere containing higher than normal amounts of nitric and sulfuric acids. Distilled water (water without any dissolved constituents), which contains no carbon dioxide, has a neutral pH of 7. Rain naturally has a slightly acidic pH of 5.6, because carbon dioxide and water in the air react together to form carbonic acid, a very weak acid. Around Washington, D.C., however, the average rain pH is between 4.2 and 4.4. Since pH is on a log scale dropping by 1 (the difference between normal rain water and acid rain) has a dramatic effect on the strength of the acid. In the United States, roughly 2/3 of all SO_2 and 1/4 of all NO_3 come from electric power generation that relies on burning fossil fuels, like coal.

Phosphorus Cycle

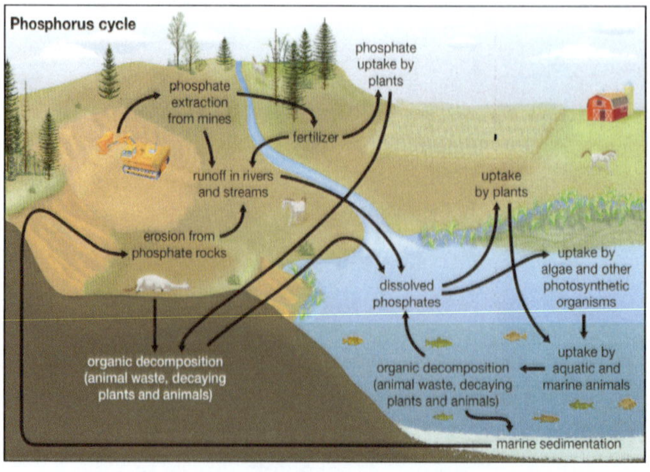

The phosphorus cycle refers to the biogeochemical cycle by which phosphorous moves through the biosphere, hydrosphere, and lithosphere. The atmosphere doesn't play a substantial role in the cycling of phosphorous. This is because phosphorous and phosphorous-based compounds cannot be found in the air in the gas state.

They are normally solids at the standard pressure and temperature range found on Earth. So, phosphorus can mainly be found as tiny dust particles in the atmosphere. It mainly cycles through the soil, water, and sediments.

That means there is a land-based phosphorus cycle and water-based phosphorus cycle. Land-based cycle transfers phosphorus from soil to plants, to animals. It's then cycled back to soil again. Water-based cycle, on the other hand, circulates phosphorous among the organisms living in the aquatic environment.

Steps of Phosphorus Cycle

Phosphorus Cycle goes through 4 main steps:

- Weathering;
- Absorption by Plants;
- Absorption by Animals;
- Return to the environment through decomposition.

Weathering

In light of the fact that phosphorous is most commonly found in rocks, the phosphorus cycle starts in the Earth's crust. Weathering causes phosphate salts to break from the rocks. The salts wash away into the ground, where they get mixed in the soil. That takes us to the second step where phosphorous is taken up by plants and animals.

Absorption by Flora and Fauna

Plants absorb phosphate salts dissolved in soil water. It is worth noting that the quantities of phosphorus in the soil are normally small. This makes it one of the main factors that undermine plant growth.

That explains why we often apply phosphate fertilizers on agricultural land. Aquatic plants absorb inorganic form phosphorous from water and bottom layer of water bodies. In light of the fact that phosphate salts are not very water-soluble, they are also limiting factors for growth of plants in aquatic ecosystems.

Absorption by Animals

Animals absorb phosphorous by eating plants or plant-eating animals. Namely, both marine and terrestrial plants are consumed by herbivorous animals, birds, and fish. Consequently, the organic form of phosphorus is transferred to the next level of consumers. Hence, the carnivores obtain phosphorous by eating herbivores.

It is worth noting that the rate of the phosphorus cycle is much faster in plants and animals than it is in rocks and sediments.

Return of Phosphorous Back to the Ecosystem

When plants, animals, birds, and fish die, they are deposited in soil and water bodies. Under conducive environmental conditions, bacteria and other microbes decompose the dead organisms. During decomposition, the organic form of phosphorous is converted into an inorganic form, which is then recycled to the soil and water.

After that, the element will end up in rock formations or sediments, where it will remain for millions of years. In the end, phosphorous is released again to the soil through weathering and absorbed by the plants. And the phosphorus cycle starts over.

Phosphorous stays in a particular ecosystem unit for a longer period, but it eventually recycles back to the environment. Hence, the percentage composition of this element remains at a virtually constant value.

Human Impact on the Phosphorus Cycle

Humans have had a significant impact on the phosphorus cycle due to a variety of human activities, such as the use of fertilizer, the distribution of food products, and artificial eutrophication. Fertilizers containing phosphorus add to the phosphorus levels in the soil and are particularly detrimental when such products are washed into local aquatic ecosystems. When phosphorus is added to waters at a rate typically achieved by natural processes, it is referred to as natural eutrophication. A natural supply of phosphorus over time provides nutrients to the water and serves to increase the productivity of that particular ecosystem. However, when foods are shipped from farms to cities, the substantial levels of Phosphorus that is drained into the water systems is called artificial or anthropogenic eutrophication. When levels of phosphorus are too high, the overabundance of plant nutrients serves to drive the excessive growth of algae. However, these algae die or form algae blooms, which are toxic to the plants and animals in the ecosystem. Thus, human activities serve to harm aquatic ecosystems, whenever excess amounts of phosphorus are leached into the water.

Phosphorus in the Environment

Ecological Function

Phosphorus is an essential nutrient for plants and animals. Phosphorus is a limiting nutrient for aquatic organisms. Phosphorus forms parts of important life-sustaining molecules that are very common in the biosphere. Phosphorus does enter the atmosphere in very small amounts when dust is dissolved in rainwater and seaspray, but remains mostly on land and in rock and soil minerals. Eighty percent of the mined phosphorus is used to make fertilizers. Phosphates from fertilizers, sewage and detergents can cause pollution in lakes and streams. Over-enrichment of phosphate in both fresh and inshore marine waters can lead to massive algae blooms which, when they die and decay, leads to eutrophication of fresh waters only. An example of this is the Canadian Experimental Lakes Area. These freshwater algal blooms should not be confused with those in saltwater environments. Recent research suggests that the predominant pollutant responsible for algal blooms in salt water estuaries and coastal marine habitats is nitrogen.

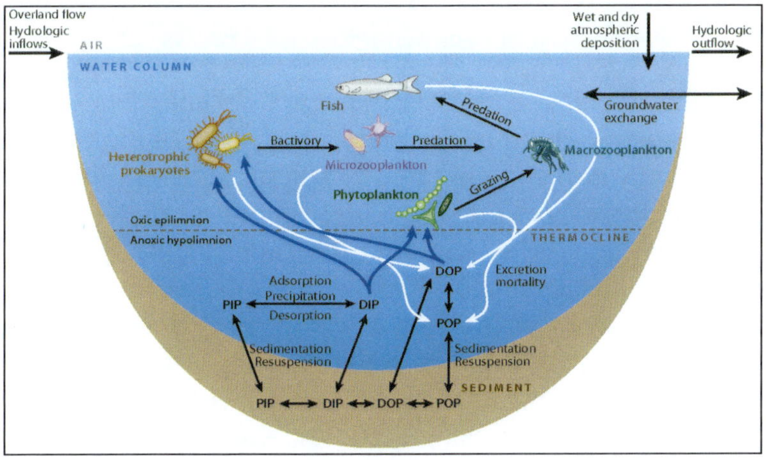

The aquatic phosphorus cycle.

Phosphorus occurs most abundantly in nature as part of the orthophosphate ion $(PO_4)^{3-}$, consisting of a P atom and 4 oxygen atoms. On land most phosphorus is found in rocks and minerals. Phosphorus-rich deposits have generally formed in the ocean or from guano, and over time, geologic processes bring ocean sediments to land. Weathering of rocks and minerals release phosphorus in a soluble form where it is taken up by plants, and it is transformed into organic compounds. The plants may then be consumed by herbivores and the phosphorus is either incorporated into their tissues or excreted. After death, the animal or plant decays, and phosphorus is returned to the soil where a large part of the phosphorus is transformed into insoluble compounds. Runoff may carry a small part of the phosphorus back to the ocean. Generally with time (thousands of years) soils become deficient in phosphorus leading to ecosystem retrogression.

Biological Function

The primary biological importance of phosphates is as a component of nucleotides, which serve as energy storage within cells (ATP) or when linked together, form the nucleic acids DNA and RNA. The double helix of our DNA is only possible because of the phosphate ester bridge that binds the helix. Besides making biomolecules, phosphorus is also found in bone and the enamel of mammalian teeth, whose strength is derived from calcium phosphate in the form of hydroxyapatite. It is also found in the exoskeleton of insects, and phospholipids (found in all biological membranes). It also functions as a buffering agent in maintaining acid base homeostasis in the human body.

Phosphorus Cycling

Phosphates move quickly through plants and animals; however, the processes that move them through the soil or ocean are very slow, making the phosphorus cycle overall one of the slowest biogeochemical cycles.

The global phosphorus cycle includes four major processes: (i) tectonic uplift and exposure of phosphorus-bearing rocks such as apatite to surface weathering; (ii) physical erosion, and chemical and biological weathering of phosphorus-bearing rocks to provide dissolved and particulate phosphorus to soils, lakes and rivers; (iii) riverine and subsurface transportation of phosphorus to various lakes and run-off to the ocean; (iv) sedimentation of particulate phosphorus (e.g., phosphorus associated with organic matter and oxide/carbonate minerals) and eventually burial in marine sediments (this process can also occur in lakes and rivers).

In terrestrial systems, bioavailable P ('reactive P') mainly comes from weathering of phosphorus-containing rocks. The most abundant primary phosphorus-mineral in the crust is apatite, which can be dissolved by natural acids generated by soil microbes and fungi, or by other chemical weathering reactions and physical erosion. The dissolved phosphorus is bioavailable to terrestrial organisms and plants and is returned to the soil after their death. Phosphorus retention by soil minerals (e.g., adsorption onto iron and aluminum oxyhydroxides in acidic soils and precipitation onto calcite in neutral-to-calcareous soils) is usually viewed as the most important factor in controlling terrestrial P-bioavailability. This process can lead to the low level of dissolved phosphorus concentrations in soil solution. Various physiological strategies are used by plants and microorganisms for obtaining phosphorus from this low level of phosphorus concentration.

Soil phosphorus is usually transported to rivers and lakes and can then either be buried in lake sediments or transported to the ocean via river runoff. Atmospheric phosphorus deposition is another important marine phosphorus source to the ocean. In surface seawater, dissolved inorganic phosphorus, mainly orthophosphate $(PO_4)^{3-}$, is assimilated by phytoplankton and transformed into organic phosphorus compounds. Phytoplankton cell lysis releases cellular dissolved inorganic and organic phosphorus to the surrounding environment. Some of the organic phosphorus compounds can be hydrolyzed by enzymes synthesized by bacteria and phytoplankton and subsequently assimilated. The vast majority of phosphorus is remineralized within the water column, and approximately 1% of associated phosphorus carried to the deep sea by the falling particles is removed from the ocean reservoir by burial in sediments. A series of diagenetic processes act to enrich sediment pore water phosphorus concentrations, resulting in an appreciable benthic return flux of phosphorus to overlying bottom waters. These processes include (i) microbial respiration of organic matter in sediments, (ii) microbial reduction and dissolution of iron and manganese (oxyhydr)oxides with subsequent release of associated phosphorus, and (iii) abiotic reduction of iron (oxyhydr)oxides by hydrogen sulfide and liberation of iron-associated phosphorus. Additionally, (i) phosphate associated with calcium carbonate and (ii) transformation of iron oxide-bound phosphorus to vivianite play critical roles in phosphorus burial in marine sediments. These processes are similar to phosphorus cycling in lakes and rivers.

Although orthophosphate $(PO_4)^-$, the dominant inorganic P species in nature, is oxidation state (P_5^+), certain microorganisms can use phosphonate and phosphite (P_3^+ oxidation state) as a P source by oxidizing it to orthophosphate. Recently, rapid production and release of reduced phosphorus compounds has provided new clues about the role of reduced P as a missing link in oceanic phosphorus.

Phosphatic Minerals

The availability of phosphorus in an ecosystem is restricted by the rate of release of this element during weathering. The release of phosphorus from apatite dissolution is a key control on ecosystem productivity. The primary mineral with significant phosphorus content, apatite $[Ca_5(PO_4)_3OH]$ undergoes carbonation.

Little of this released phosphorus is taken up by biota (organic form), whereas a larger proportion reacts with other soil minerals. This leads to precipitation into unavailable forms in the later stage of weathering and soil development. Available phosphorus is found in a biogeochemical cycle in the upper soil profile, while phosphorus found at lower depths is primarily involved in geochemical reactions with secondary minerals. Plant growth depends on the rapid root uptake of phosphorus released from dead organic matter in the biochemical cycle. Phosphorus is limited in supply for plant growth. Phosphates move quickly through plants and animals; however, the processes that move them through the soil or ocean are very slow, making the phosphorus cycle overall one of the slowest biogeochemical cycles.

Low-molecular-weight (LMW) organic acids are found in soils. They originate from the activities of various microorganisms in soils or may be exuded from the roots of living plants. Several of those organic acids are capable of forming stable organo-metal complexes with various metal ions found in soil solutions. As a result, these processes may lead to the release of inorganic phosphorus associated with aluminum, iron, and calcium in soil minerals. The production and release of oxalic acid by mycorrhizal fungi explain their importance in maintaining and supplying phosphorus to plants.

The availability of organic phosphorus to support microbial, plant and animal growth depends on the rate of their degradation to generate free phosphate. There are various enzymes such as phosphatases, nucleases and phytase involved for the degradation. Some of the abiotic pathways in the environment studied are hydrolytic reactions and photolytic reactions. Enzymatic hydrolysis of organic phosphorus is an essential step in the biogeochemical phosphorus cycle, including the phosphorus nutrition of plants and microorganisms and the transfer of organic phosphorus from soil to bodies of water. Many organisms rely on the soil derived phosphorus for their phosphorus nutrition.

Phosphorus and Eutrophication

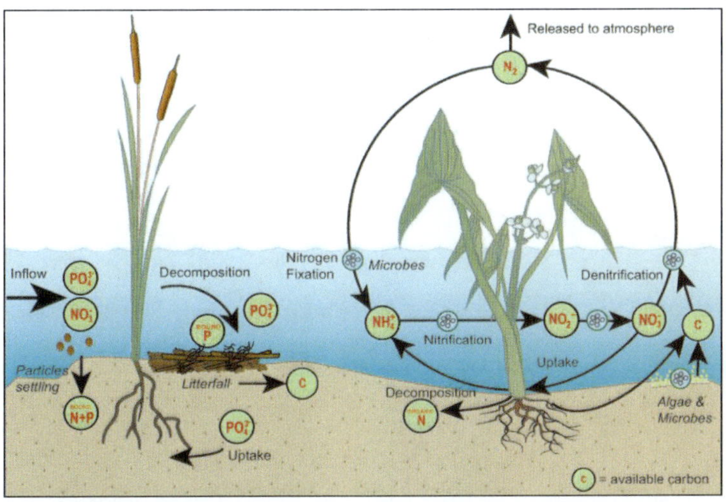

A simplified illustration of the nitrogen and phosphorus cycles in a wetland.

Eutrophication is an enrichment of water by nutrient that lead to structural changes to the aquatic ecosystem such as algae bloom, deoxygenation, and reduction of fish species. The primary source that contributes to the eutrophication is considered as nitrogen and phosphorus. When these two elements exceed the capacity of the water body, eutrophication occurs. Phosphorus that enters lakes will accumulate in the sediments and the biosphere; it also can be recycled from the sediments and the water system. Drainage water from agricultural land also carries phosphorus and nitrogen. Since a large amount of phosphorus is in the soil contents, so the overuse of fertilizers and over-enrichment with nutrients will lead to increasing the amount of phosphorus concentration in agricultural runoff. When eroded soil enters the lake, both phosphorus and the nitrogen in the soil contribute to eutrophication, and erosion caused by deforestation which also results from uncontrolled planning and urbanization.

Wetland

Wetlands are frequently applied to solve the issue of eutrophication. Nitrate is transformed in wetlands to free nitrogen and discharged to the air. Phosphorus is adsorbed by wetland soils which are taken up by the plants. Therefore, wetlands could help to reduce the concentration of nitrogen and phosphorus to remit and solve the eutrophication. However, wetland soils can only hold a limited amount of phosphorus. To remove phosphorus continually, it is necessary to add more new soils within the wetland from remnant plant stems, leaves, root debris, and undecomposable parts of dead algae, bacteria, fungi, and invertebrates.

Hydrogen Cycle

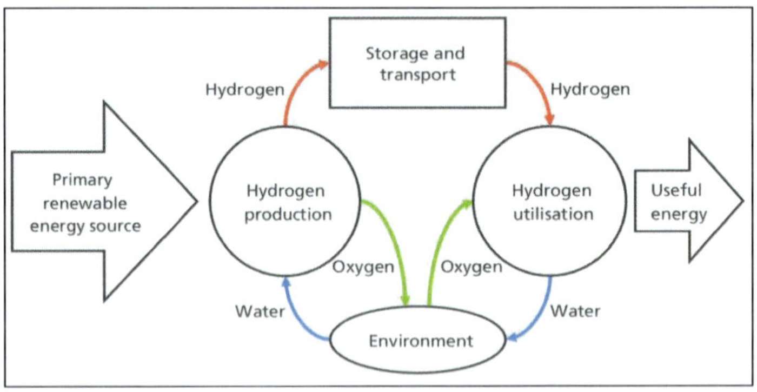

The hydrogen cycle Is that process in which the hydrogen moves by means of the water around the earth, being thus an essential part of the chemical and atomic composition of this element.

The hydrosphere obtains hydrogen only from water, an element formed exclusively by the combination of oxygen and hydrogen. During photographic synthesis, hydrogen is produced by the dissociation of water that forms glucose after combing with carbon dioxide.

Plants provide food for herbivores and these animals get glucose and plant proteins only. Hydrogen forms carbohydrates, which is an important source of energy for the living, and these carbohydrates come as food.

There are innumerable types of living beings on earth. All these are basically composed of carbon, nitrogen, Oxygen and hydrogen. Animals obtain these elements of nature and processes such as formation, growth and decomposition normally occur in it.

Several cycles occur as a result of each of these processes, and because of them, they are linked together, establishing a balance.

Phases of the Hydrogen Cycle

Hydrogen atoms can be stored as a gas or a high pressure liquid. Hydrogen is often stored as liquid hydrogen because it takes up less space than hydrogen in its normal gas form.

When a hydrogen atom joins a strongly electronegative atom that exists in the vicinity of another electronegative atom with a solitary pair of electrons, it makes a hydrogen bond, which forms a molecule. Two hydrogen atoms form a hydrogen molecule, H_2 for short.

Hydrogen is a key component of many biogeochemical cycles, including the water cycle, the carbon cycle, the nitrogen cycle and the sulfur cycle. Because hydrogen is a component of the water molecule, the hydrogen cycle and the water cycle are deeply linked.

Plants also recombine water and soil carbon dioxide and atmosphere to form glucose in a process known as photosynthesis. If the plant is consumed, the hydrogen molecules are transferred to the grazing animal.

Organic matter is stored in floors as the plant or animal dies, and the hydrogen molecules are released back into the atmosphere by oxidation.

1. Evaporation: Most of the hydrogen in our planet is in the water, so the hydrogen cycle is closely related to the hydrological cycle. The hydrogen cycle begins with the evaporation of the water surface.
2. Condensation: The hydrosphere includes the atmosphere, land, surface water and groundwater. As the water moves through the cycle, the state changes between the liquid, solid and gas phases.

   Water moves through different reservoirs, including ocean, atmosphere, groundwaters, rivers and glaciers, by physical evaporation processes (including plant transpiration), sublimation, precipitation, infiltration, runoff and Sub surface flux.
3. Transpiration: Plants absorb soil water through their roots and then pump it and supply nutrients to their leaves. The transpiration represents approximately 10% of the evaporated water.

   This is the discharge of water vapor from the leaves of plants into the atmosphere. It is a process that the eye cannot see, even though the amounts of moisture it implies are significant. It is believed that a large oak can transpire 151,000 liters per year.

   Perspiration is also the reason why there is more humidity in places with a lot of vegetation cover. The amount of water that transpires through this process depends on the plant itself, the moisture in the soil (soil), the surrounding temperature and the movement of the wind around the plant.
4. Precipitation: It is the fall of water in any form to the land that gives way to infiltration, which is the process in which water is absorbed in the soil or flows through the surface. This process is repeated over and over again as part of the terrestrial cycles Renewable resources.

Function of Hydrogen on Earth

- It is mainly used to create Water. Hydrogen gas can be used for the reduction of metallic ore.
- Chemical industries also use it for the production of hydrochloric acid. The same hydrogen gas is required for atomic hydrogen (AHW) welding.
- There are a variety of uses for hydrogen. It is the lightest element and can be used as a lifting agent on balloons, although it is also highly flammable, so it can be dangerous. This property and others make the hydrogen suitable for use as fuel.
- Since hydrogen is highly flammable, especially when mixed with pure oxygen, it is used as fuel in rockets. These, usually, combine liquid hydrogen with liquid oxygen to make an explosive mixture.
- Hydrogen is one of the cleanest fuels because igniting the result is simple water. This is one of the main reasons why there are efforts to create engines that can be powered by the use of this gas.
- Although hydrogen is highly flammable, so is gasoline. Although care must be taken, the amount of hydrogen used in a car would present no more danger than the amount of gasoline used.
- Despite being one of the cleanest fuels on the planet, its high cost for mass production makes it impossible in the near future to use it for cars for commercial and domestic use.
- When hydrogen is heated to extreme temperatures, the nuclei of its atoms will merge to create helium nuclei. This fusion results in the release of a huge amount of energy, called thermonuclear energy. This process is what creates the solar energy.
- Electric generators use gas as a refrigerant, which has led many plants to use it as a leak-checking agent. Other applications include processing and production of ammonia.
- Ammonia is part of many household cleaning products. It is also a hydrogenation agent used to switch unsaturated unsaturated fats to saturated oils and fats.

Null Cycle

In atmospheric chemistry, a null cycle is a catalytic cycle that simply interconverts chemical species without leading to net production or removal of any component. In the stratosphere, null cycles and when the null cycles are broken are very important to the ozone layer.

One of the most important null cycles takes place in the stratosphere, with the photolysis of ozone by photons with wavelengths less than 330 nanometers. This photolysis produces a monatomic oxygen that then reacts with the diatomic oxygen producing ozone. There is no net molecular or atomic change; however, this reaction converts the energy from the photons into heat energy warming the stratosphere.

$$O + hv\ (\lambda\ 1D)$$
$$O\ (1D) + M \rightarrow O\ (3P) + M$$
$$O\ (3P) + O \rightarrow O$$
$$\text{Net}: hv \rightarrow H$$

The null cycle can be broken in the presence of certain molecules, leading to a net increase or decrease in ozone in the stratosphere. One important example is NO emissions into the stratosphere. The NO reacts with both the atomic oxygen and ozone leading to a net decrease in ozone. This is particularly important at night when NO cannot photolyze.

$$NO + O \rightarrow NO + O$$
$$NO + O(1D) \rightarrow NO + O$$
$$\text{Net}: O + O(1D) \rightarrow 2O\ (\text{net loss of ozone})$$

Null cycles can also occur in the troposphere. One example is the null cycle that occurs during the day between NO and ozone.

Tropospheric Null Cycle

$$O + NO \rightarrow O + NO$$
$$NO + hv \rightarrow NO + O(3P)$$
$$O\ (3P) + O + M \rightarrow O + M$$
$$\text{Net}: hv \rightarrow H$$

This cycle links ozone to NOx in the troposphere during daytime. In equilibrium, described by the Leighton relationship, solar radiation and the NO: NO ratio determine ozone abundance, maximizing around noon time.

References

- Atmospheric-science: planetary-science.org, Retrieved 2 May, 2019
- Atmospheric-science, science: britannica.com, Retrieved 18 January, 2019
- Madigan MT, Martino JM (2006). Brock Biology of Microorganisms (11th ed.). Pearson. p. 136. ISBN 978-0-13-196893-6
- Layers-of-the-atmosphere, geophysical: lumenlearning.com, Retrieved 10 April, 2019
- Atmospheric-temperature, science: encyclopedia.com, Retrieved 14 June, 2019
- Moir, JWB (editor) (2011). Nitrogen Cycling in Bacteria: Molecular Analysis. Caister Academic Press. ISBN 978-1-904455-86-8
- Temperature, climate-meteorology, science: britannica.com, Retrieved 4 February, 2019
- Atmospheric-temperature-range: definitions.net, Retrieved 22 May, 2019
- Upper-air-temperature: remss.com, Retrieved 25 July, 2019

- Lalli, Parsons, Carol, Timothy (1997). Biological oceanography: An introduction. Butterworth-Heinemann. ISBN 978-0-7506-3384-0
- Atmosphere-Composition-Structure: jrank.org, Retrieved 5 March, 2019
- Carbon-cycle, biogeochemical-cycles, air-climate-weather: enviroliteracy.org, Retrieved 28 February, 2019
- Process-of-oxygen-cycle, environment: eartheclipse.com, Retrieved 9 April, 2019
- Sutton, Mark A., editor (2011-04-14). The European nitrogen assessment: sources, effects, and policy perspectives. ISBN 9781107006126. OCLC 690090202
- Sulfur-cycle, biogeochemical-cycles: enviroliteracy.org, Retrieved 19 March, 2019
- Brimblecombe, Peter (2014). "The global sulfur cycle". Treatise on Geochemistry. 10. Amsterdam: Elsevier. pp. 559–591. doi:10.1016/B978-0-08-095975-7.00814-7. ISBN 9780080983004
- Phosphorus-cycle-definition-steps-facts: eartheclipse.com, Retrieved 11 August, 2019
- The-hydrogen-cycle-and-its-most-important-phases: lifepersona.com, Retrieved 12 January, 2019

4

Weather and Weather Forecasting

Weather refers to the state of the atmosphere which describes the extent of how hot, cold, wet, dry, stormy, clear or cloudy it is. Applying science and technology to predict the conditions of the atmosphere at a specific location and time is called weather forecasting. The topics elaborated in this chapter will help in gaining a better perspective about the diverse aspects of weather and weather forecasting.

Weather

Weather is the mix of events that happen each day in our atmosphere. Weather is different in different parts of the world and changes over minutes, hours, days and weeks. Most weather happens in the troposphere, the part of Earth's atmosphere that is closest to the ground.

The weather events happening in an area are controlled by changes in air pressure. Air pressure is caused by the weight of the huge numbers of air molecules that make up the atmosphere. Typically, when air pressure is high there skies are clear and blue. The high pressure causes air to flow down and fan out when it gets near the ground, preventing clouds from forming. When air pressure is low, air flows together and then upward where it converges, rising, cooling, and forming clouds. Remember to bring an umbrella with you on low pressure days because those clouds might cause rain or other types of precipitation.

Meteorologists develop local or regional weather forecasts including predictions for several days into the future. The best forecasts take into account the weather events that are happening over a broad region. Knowing where storms are now can help forecasters predict where storms will be tomorrow and the next day. Technology, such as weather satellites and Doppler radar, helps the process of looking over a large area, as does the network of weather observations.

The chaotic nature of the atmosphere means that it will probably always be impossible to predict the weather more than two weeks ahead; however, new technologies combined with more traditional methods are allowing forecasters to develop better and more complete forecasts.

The average weather pattern in a place over several decades is called climate. Different regions have different regional climates. For example, the climate of Antarctica is quite different than the climate of a tropical island. Global climate refers to the average of all regional climates.

As global climate changes, weather patterns are expected to change as well. While it is impossible to say whether a particular day's weather was affected by climate change, it is possible to predict how patterns might change. For example, scientists predict more severe weather events as climate warms. Also, they predict more hot summer days and fewer extreme cold winter days. That doesn't mean that there will be no more winter weather, in fact, large snowstorms might even be more likely in some areas as less cold air is able to carry more water with which to make snowflakes.

Weather is also affected by climate events like El Nino and La Nina (together known as ENSO). Climate events like these affect the weather in many areas of the world causing extreme events like storms and droughts.

Condensation

Condensation is the process where water vapor becomes liquid. It is the reverse of evaporation, where liquid water becomes a vapor.

Condensation happens one of two ways: Either the air is cooled to its dew point or it becomes so saturated with water vapor that it cannot hold any more water.

Dew Point

Dew point is the temperature at which condensation happens. (Dew is simply condensed water in the atmosphere.) Air temperatures can reach or fall below the dew point naturally, as they often do at night. That's why lawns, cars, and houses are often coated with water droplets in the morning.

Condensation can also produce water droplets on the outside of soda cans or glasses of cold water. When warm air hits the cold surface, it reaches its dew point and condenses. This leaves droplets of water on the glass or can.

When a pocket of air becomes full of water vapor, clouds form. The point at which condensation starts can be easily viewed in cumulus clouds, which have flat bottoms. Those flat bottoms are where vapor begins to condense into water droplets.

Saturation

Clouds are simply masses of water droplets in the atmosphere. Molecules in water vapor are far apart from one another. As more water vapor collects in clouds, they can become saturated with water vapor. Saturated clouds cannot hold any more water vapor. When clouds are saturated with water vapor, the density, or closeness, of the molecules increases. The vapor condenses and becomes rain.

Cold air holds less water vapor than warm air. This is why warm climates are often more humid than cold ones: Water vapor remains in the air instead of condensing into rain. Cold climates are more likely to have rain, because water vapor condenses more easily there.

From the outside in: condensation, can, cola.

Weather Forecasting

Weather forecasting entails predicting how the present state of the atmosphere will change. Present weather conditions are obtained by ground observations, observations from ships, observation from aircraft, radio sounds, doppler radar and satellites. This information is sent to meteorological centers where the data are collected, analyzed and made into a variety of charts, maps and graphs. Modern high-speed computers transfer the many thousands of observations onto surface and upper-air maps.

Weather forecasts provide critical information about future weather. There are various techniques involved in weather forecasting, from relatively simple observation of the sky to highly complex computerized mathematical models. Weather prediction could be one day/one week or a few months ahead. The accuracy of weather forecasts however, falls significantly beyond a week. Weather forecasting remains a complex business, due to its chaotic and unpredictable nature. It remains a process that is neither wholly science nor wholly art. It is known that persons with little or no formal training can develop considerable forecasting skill. For example, farmers often are quite capable of making their own short term forecasts of those meteorological factors that directly influence their livelihood, and a similar statement can be made about pilots, fishermen, mountain climbers, etc. Weather phenomena, usually of a complex nature, have a direct impact on the safety and/or economic stability of such persons. Accurate weather forecast models are important to third world countries, where the entire agriculture depends upon weather. It is thus a major concern to identify any trends for weather parameters to deviate from its periodicity, which would disrupt the economy of the country. This fear has been aggravated due to threat by the global warming and greenhouse effect. The impact of extreme weather phenomena on society is growing more and more costly, causing infrastructure damage, injury and the loss of life.

As practiced by the professionally trained meteorologist, weather forecasting today is a highly developed skill that is grounded in scientific principle and method and that makes use of advanced technological tools. The notable improvement in forecast accuracy that has been achieved since 1950 is a direct outgrowth of technological developments, basic and applied research, and the application of new knowledge and methods by weather forecasters. High-speed computers, meteorological satellites, and weather radars are tools that have played major roles in improving weather forecasts.

Several other factors have contributed significantly to this increase in forecasting accuracy. One is the development of statistical methods for enhancing the scope and accuracy of model predictions. Another is the improved observational capability afforded by meteorological satellites. A third primary reason for the increase in accuracy is the continued improvement of the initial conditions prepared for the forecast models. Statistical methods allow a wider variety of meteorological elements to be predicted than do the models alone, and they tailor the geographically less precise model forecasts to specific locations. Satellites now provide the capability for nearly continuous viewing and remote sensing of the atmosphere on a global scale. The improvement in initial conditions is the result of an increased number of observations and better use of the observations in computational techniques.

Traditional Weather Forecasting

Weather forecasting is the application of science and technology to predict the state of the atmosphere for a given location. Weather forecasts are made by collecting quantitative data about the current state of the atmosphere and using scientific understanding of atmospheric processes to project how the atmosphere will evolve. There are a variety of end users to weather forecasts. Weather warnings are important forecasts because they are used to protect life and property.

In ancient times, forecasting was mostly based on weather pattern observation. Over the years, the study of weather patterns has resulted in various techniques for rainfall forecasting. Present rainfall forecasting embodies a combination of computer models, interpretation, and an acquaintance of weather patterns. The following technique was used for existing weather prediction:

Use of a Barometer

Measurements of barometric pressure and the pressure tendency have been used in forecasting since the late 19th century. The larger the change in pressure, the larger the change in weather can be expected. If the pressure drop is rapid, a low pressure system is approaching, and there is a greater chance of rain.

Looking at the Sky

Along with pressure tendency, the condition of the sky is one of the most important parameters used to forecast weather in mountainous areas. Thickening of cloud cover or the invasion of a higher cloud deck is an indication of rain in the near future. At night, high thin clouds can lead to halos around the moon, which indicates the approach of a warm front and its associated rain. Morning fog portends fair conditions, as rainy conditions are preceded by wind or clouds which prevent fog formation.

Nowcasting

The forecasting of the weather within the next six hours is often referred to as nowcasting. In this time range, it is possible to forecast smaller features such as individual showers and thunderstorms with reasonable accuracy, as well as other features too small to be resolved by a computer model. A human, given the latest radar, satellite and observational data will be able to make a better analysis of the small scale features present and so will be able to make a more accurate forecast for the following few hours.

Analog Technique

The analog technique is a complex way of making a forecast, requiring the forecaster to remember a previous weather event which is expected to be mimicked by an upcoming event. It remains a useful method of observing rainfall in places such as oceans, as well as the forecasting of precipitation amounts and distribution in the future. A similar technique is used in medium range forecasting, which is known as teleconnections, when systems in other locations are used to help pin down the location of another system within the surrounding regime.

Numerical Weather Prediction Model

Numerical Weather Prediction (NWP) is the science of predicting the weather using models of the atmosphere and computational techniques. Current weather conditions are used at the input of the mathematical models of the atmosphere to predict the weather. This model usually provides surrounding point around the wind farm with a spatial resolution of a few kilometers. NWP uses the power of computers to make a forecast. A forecaster examines how the features predicted by the computer will interact to produce the day's weather. The NWP method is flawed in that the equations used by the models to simulate the atmosphere are not precise.

A number of weather forecasting agencies operate modeling centers where supercomputers are used to run NWP models that span the entire globe. These include the National Center for Environmental Prediction (NCEP) in the United States, the United Kingdom Meteorological Office (UKMO), and the European Centre for Medium-range Weather Forecasts (ECMWF). Although costly, a global approach to NWP is essential, especially for long-range forecasting. For this reason, achieving accurate forecasts requires an accurate analysis from which to get the model started. This involves a computer-based process called data assimilation, in which the most recent weather observations from around the world are combined with model forecasts to create a global analysis of current conditions. This becomes the starting point for the next run of the NWP model, and is the computer equivalent of the manual analysis cycle that forecasters carry out on an on-going basis. Global models play a key role in modern weather forecasting, and meteorologists at Met Service routinely use the NCEP, UKMO and ECMWF models to assist with day-to-day production of forecasts and weather warnings. These models give insight into the behavior of weather systems on a large scale, without much emphasis on local detail.

Ensemble Forecasting

To predict the weather forecast meteorologists have developed atmospheric models that approximate the atmosphere by using ensemble forecasting to describe how atmospheric temperature, pressure and moisture will change over time. The equations are programmed into a computer and the data on the present atmospheric conditions are fed into the computer. The computer solves the equations to determine how the different atmospheric variables will change over the next few minutes. The computer repeats this procedure again and again using the output from one cycle as the input for the next cycle. For some desired time in the future, the computer prints its calculated information. It then analyzes the data, drawing the lines for the projected position of the various pressure systems. A forecaster uses the prognostic chart as a guide to predicting the weather. There are many atmospheric models that represent the atmosphere, with each one interpreting the atmosphere in a slightly different way. Weather forecasts made for 12 and 24 hours are typically accurate. Forecasts made for two or three days are usually good. Beyond above five days, forecast accuracy falls off rapidly. Weather information can also come from remote sensing, particularly radar and satellites.

Radar

Radar stands for Radio Detection and Ranging. In radar, a transmitter sends out radio waves. The radio waves bounce off the nearest object and then return to a receiver. Weather radar can sense many characteristics of precipitation, its location, motion, intensity, and the likelihood of future precipitation. Most weather radar is Doppler radar, which can also track how fast the precipitation falls. Radar can outline the structure of a storm and in doing so estimates the possibility that it will produce severe weather condition.

Weather Satellites

Weather satellites have been increasingly important sources of weather data since the first one was launched in 1952. Weather satellites are the best way to monitor large scale systems, like storms. Satellites can also monitor the spread of ash from a volcanic eruption, smoke from fires, and pollution. They are able to record long-term changes. Figure shows one of the geostationary satellites that monitors conditions over the world. Weather satellites may observe all energy from all wavelengths in the electromagnetic spectrum. Most important are the visible light and infrared (heat) frequencies.

Geostationary satellites.

Weather Maps

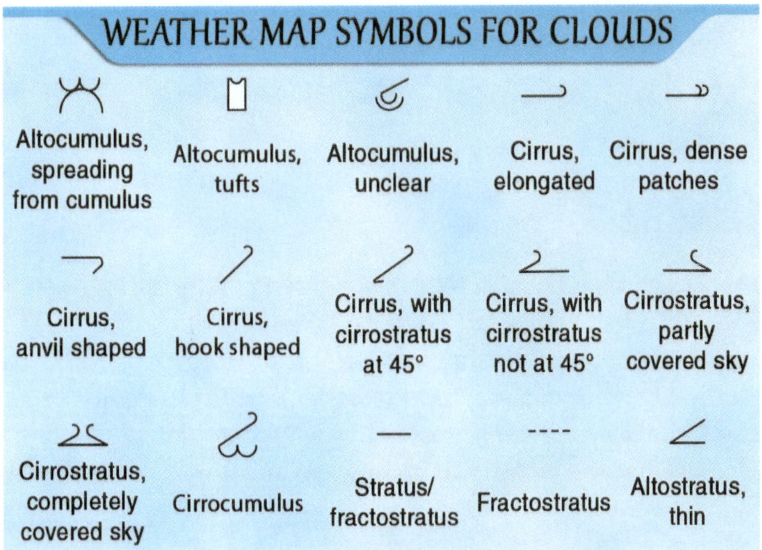

Weather maps simply and graphically depict meteorological conditions in the atmosphere. Weather maps may display only one feature of the atmosphere or multiple features. They can depict information from computer models or from human observations. Weather maps are found in newspapers, on television, and on the Internet.

On a weather map, each weather station will have important meteorological conditions plotted. These conditions may include temperature, current weather, dew point, cloud cover, sea level air pressure, wind speed and direction. On a weather map, meteorologists use many different symbols. These symbols give them a quick and easy way to put information onto the map. Figure shows some of these symbols used for Weather Map.

Generally, there are two methods used in weather forecasting namely the empirical and dynamical approach. Occurrence of analogues and is often referred to by meteorologists as analogue forecasting. This approach normally is useful for predicting local recorded cases are plentiful. Dynamical approach is based upon equation and forward simulations of the atmosphere and is often referred to as computer modeling. This dynamical approach is useful for modeling large predict short-term weather efficiently. Most of the weather forecasting systems combine both empirical and dynamical approaches.

Types of Weather Forecasting

A daily weather forecast involves the work of thousands of observers and meteorologists all over the world. Modern computers make forecasts more accurate than ever, and weather satellites orbiting the earth take photographs of clouds from space.

Forecasters use the observations from ground and space, along with formulas and rules based on experience of what has happened in the past, and then make their forecast.

Meteorologists actually use a combination of several different methods to come up with their daily weather forecasts. They are:

a. Persistence Forecasting,

b. Synoptic Forecasting,

c. Statistical Forecasting,

d. Computer forecasting.

Persistence Forecasting

The simplest method of forecasting the weather is persistence forecasting. It relies upon today's conditions to forecast the conditions tomorrow. This can be a valid way of forecasting the weather when it is in a steady state, such as during the summer season in the tropics. This method of forecasting strongly depends upon the presence of a stagnant weather pattern. It can be useful in both short range forecasts and long range forecasts. This assumes that what the weather is doing now is what it will continue to do. To find out what the weather is doing, meteorologists make weather observations.

Synoptic Forecasting

This method uses the basic rules for forecasting. Meteorologists take their observations, and apply those rules to make a short-term forecast.

Statistical Forecasting

Meteorologists ask themselves, what does it usually do this time of the year? Records of average temperatures, average rainfall and average snowfall over the years give forecasters an idea of what the weather is "supposed to be like" at a certain time of the year.

Computer Forecasting

Forecasters take their observations and plug the numbers into complicated equations. Several ultra-high-speed computers run these various equations to make computer "models" which give a forecast for the next several days. Often, different equations produce different results, so meteorologists must always use the other forecasting methods along with this one.

Using all the above methods, forecasters come up with their "best guess" as to what weather conditions will be over the next few days.

Weather forecasting now has a wide range of operational products that traditionally are classified under the following groups:

- Very short-range forecast,
- Short-range forecast,
- Medium-range forecast,
- Long-range forecast.

Each weather forecast can be defined on the basis of the following criteria:

- Dominant technology;
- Temporal range of validity after emission;
- Characteristics of input and output time and space resolution;
- Broadcasting needs;
- Accuracy.

Importance of Weather Forecasting

Weather forecasting is used in many situations like severe weather alerts and advisories, predicting the behavior of the cloud for air transport, prediction of waterways in a sea, agricultural development and avoiding forest fire.

Severe Weather Alerts and Advisories

A major part of modern weather forecasting is the severe weather alerts and advisories which are the national weather service's issue in anticipation of severe or hazardous weather are expected. This is done to protect life and property. Some of the most commonly known of severe weather advisories are the severe thunderstorm and tornado warning, as well as the severe thunderstorm and tornado watch. Other forms of these advisories include winter weather, high wind, flood, tropical cyclone, and fog. Severe weather advisories and alerts are broadcast through the media, including radio, using emergency systems as the Emergency Alert System which breaks into regular programming.

Predicting the Behavior of the Cloud for Air Transport

The aviation industry is especially sensitive to the weather and accurate weather forecasting is essential. Fog or exceptionally low ceilings can prevent many aircraft from landing and taking off.

Turbulence and icing are also significant in-flight hazards. Thunderstorms are a problem for all aircrafts because of severe turbulence due to their updrafts and outflow boundaries, icing due to the heavy precipitation, as well as large hail, strong winds, and lightning, all of which can cause severe damage to an aircraft in flight. Volcanic ash is also a significant problem for aviation, as aircraft can lose engine power within ash clouds.

Prediction of Waterways in a Sea

Commercial and recreational use of waterways can be limited significantly by wind direction, speed, wave periodicity, high tides and precipitation. These factors can each influence the safety of marine transit. Consequently, a variety of codes have been established to efficiently transmit detailed marine weather forecasts to vessel pilots via radio, for example marine forecast. Typical weather forecasts can be received at sea through the use of Radio fax.

Agricultural Development

Weather plays an important role in agricultural production. It has a profound influence on the growth, development and yields of a crop, incidence of pests and diseases, water needs and fertilizer requirements in terms of differences in nutrient mobilization due to water stresses and timeliness and effectiveness of prophylactic and cultural operations on crops. Weather aberrations may cause (i) physical damage to crops and (ii) soil erosion. The quality of the crop produced during movement from field to storage and transport to market depends on weather. Bad weather may affect the quality of the produce during transport and viability and vigor of seeds and planting material during storage.

Avoiding Forest Fire

Weather forecasting of wind, precipitations and humidity is essential for preventing and controlling wildfires. Different indices, like the Forest fire weather index and the Haines Index, have been developed to predict the areas more at risk to experience fire from natural or human causes. Conditions for the development of harmful insects can also be predicted by weather forecasting.

Military Applications

Military weather forecasters present weather conditions to the war fighter community. Military weather forecasters provide pre-flight and in-flight weather briefs to pilots and provide real time resource protection services for military installations. Naval forecasters cover the waters and ship weather forecasts. The Navy provides a special service to both themselves and the rest of the federal government by issuing forecasts for tropical cyclone across the Pacific and Indian Oceans through their Joint Typhoon Warning Center.

Air Force

Air Force Weather provides weather forecasting for the Air Force and the Army. Air Force forecasters cover air operations in both wartime and peacetime operations and provide Army support. Military and civilian forecasters actively cooperate in analyzing and creating weather forecast products.

Methods of Weather Forecasting

Of course for weather forecast to exist there must be methods on which it is done. These methods are as follows:

Persistence Forecasting

Persistence forecasting is the easiest method of forecasting which assumes a continuation of the present. It relies upon today's conditions to forecast the weather when it is steady state, such as during the summer season in the tropics. This method of forecasting strongly depends upon the presence of a stagnant weather pattern. It can be useful in both short-range forecasts and long-range forecasts.

Persistence forecasts are used by local forecasters in determining such events as the time of the arrival of a thunderstorm that is moving toward their region. Persistence forecasts do not account for changes that might occur in the intensity or in the path of a weather system, and they do not predict the formation. Because of these limitations and the rapidity with which weather system change in most geographical regions, persistence forecasts break down after twelve hours, or a day at most.

Climatology Forecasting

Whereas persistence forecasting is most accurate over short periods (before factors for change have had time to operate), the best estimate of the weather a long time ahead is the average value of past measurements there at that time of day and year.

Climatology forecast relies on the observation that weather for a particular day at a location does not change much from one year to the next. As a result, a long term average of weather on a certain day or month should be a good guess as the weather for that day or month. The most obvious climatology forecast in this part of the world (Nigeria) is, "Cold in December, warm in July (the popular July break)". One does not need to be a meteorologist to make that forecast.

Today's numerical forecast methods still use climatological statistics as a "reality check". There make sure that the computer models are not going off the deep end, climatologically speaking.

Looking at the Sky

Along with pressure tendency, use of the sky condition is one of more important weather parameters that can be used to forecast weather in mountainous areas. Thickening of cloud cover or the invasion of a higher cloud deck is indicative of rain in the near future. Morning fog portends fair conditions, as rainy conditions are preceded by wind or clouds, which prevent fog formation. The approach of a line of thunderstorm could indicate the approach of a cold front. Cloud free skies are indicative of fair weather for the near future. The use of sky cover in weather prediction has led to various weather lore over the centuries.

Use of a Barometer

Using barometric pressure and the pressure tendency (xthe change of pressure over time) has been used in forecasting since the late 19th century. The larger the change in pressure, especially, if more than 2.54mmHg, the larger the change in weather can be expected. If the pressure drop is rapid, a low-pressure system is approaching, and there is a greater chance of rain. Rapid pressure rises are associated with improving weather conditions, such as clearing skies.

Nowcasting

The forecasting of the weather within the next six hours is often referred to as nowcasting. In this time range, it is possible to forecast smaller features such as individual showers and thunderstorms with reasonable accuracy, as well as other features too small to be resolved by a computer model. A human given the latest radar, satellite and observational data will be able to make a better analysis of the small scale features present and so will be able to make a more accurate forecast for the following few hours.

Severe weather is typically short-lived (less than two hours) and, due to its mesoscale character (less than one hundred kilometers), it affects local/regional areas necessitating site-specific forecasts. Included in this category are thunderstorms, gust fronts, tornadoes, high winds especially along coasts, over lakes and mountains, heavy snow and freezing precipitation. The development of radar networks, new instruments and high speed communication links has provided a means of issuing warnings of such phenomena.

Several countries including Nigeria have recently developed integrated satellite and radar systems to provide information on the horizontal and vertical extent of thunderstorms, for example. Such data are, supplemented by networks of automatic weather stations that measure wind, temperature and humidity.

Nowcasting methods use highly automated computers and image analysis systems to integrate data from a variety of sources rapidly. Interpretation of the data displays requires skilled personnel and or extensive software to provide appropriate information. The prompt forecasting of wind shear and downburst hazards at airports is one example of the importance of nowcasting procedures.

Use of Forecasting Models

In the past, the human forecasters were responsible for generating the entire weather forecast based upon available observation. Today, human input is generally confined to choosing a model based on various parameters, such as model biases and performance. Using a consensus of forecast models, as well as ensemble members of the various models, can help reduce forecast error. However, regardless how small the average error becomes with any individual system, large errors within any particular piece of guidance are still possible on any given model run. Humans can use knowledge of local effects, which may be too small in size to be resolved by the model to add information to the forecast.

Analogue Forecasting

The analogue method is a complex way of making a forecast, requiring the forecaster to remember a previous weather event which is expected to be mimicked by an upcoming event. The analogue forecaster's task is to locate the date in history when the weather is a perfect match, or analogue, to today's weather. Then the forecast for tomorrow is simple – whatever happened in the day after the analogue will be the weather for tomorrow. The forecast for the day after tomorrow is whatever happened in the second day after the analogue, and so forth.

What makes it a difficult method to use is that, there is rarely a perfect analogue for an event in the future. In fact, no two patterns or sequences of weather are ever identical. There may, for example, be five reasonable analogues for a particular month, but examination of the succeeding weather sequences might show mild, rainy weather in two cases and cold spells in the other three. In the preparation of the forecast therefore, many factors, which can affect the weather trends, such as sea temperature and the extent or amount of rainfall, have to be taken into consideration.

Some call this type of forecasting pattern recognition. It remains a useful method of observing rainfall over data voids such as oceans, as well as the forecasting of precipitation amounts and distribution in the future. A similar method is used in medium range forecasting, which is known as teleconnections – when systems in other locations are used to help pin down the location of another system within the surrounding regime. Teleconnections are used by forecasters today to make general forecasts months into the future. While in short term forecasting, the pattern recognition is still used by weather forecasters to supplement today's computerized methods. But in the end the complexities of weather, like human personalities, defy simple categorization.

The main problem with this method may well be the lack of complete enough information, which also of course, limits the usefulness of numerical weather predictions.

Ensemble Forecasting

Although a forecast model will predict weather features evolving realistically into the distant future, the errors in a forecast will inevitably grow with time due to the chaotic nature of the atmosphere and the inexactness of the initial observations. The detail that can be given in a forecast therefore decreases with time as these errors increase. These become a point when the errors are so large that the forecast has no correlation with the actual state of the atmosphere.

However, looking at a single forecast gives no indication of how likely that forecast is to be correct. Ensemble forecasting entails the production of many forecasts in order to reflect the uncertainty into the initial state of the atmosphere (due to the errors in the observations and insufficient sampling). The uncertainty in the forecast can then be assessed by the range of different forecasts produced. However, the simple logic behind ensemble forecasting is that two runs of a model are not enough to base a forecast upon.

Ensemble forecasts are increasingly being used for operational weather forecasting. Ensemble forecasting requires a sophisticated understanding of the atmosphere and computer models.

Applications of Weather Forecast

The importance of accurate weather forecasts cannot be over emphasized as the needs for them are always craved for in virtually every aspect of life.

Severe Weather Alerts and Advisories

A major part of modern weather forecasting is the severe weather alerts and advisories, which the national weather service's issue in the case that severe or hazardous weather is expected. This is done to protect life and property. Some of the most commonly known severe weather advisories are the severe thunderstorm and tornado warnings, as well as the recent warnings about areas that are prone to flood in some part of Nigeria by the National Meteorological Agency. Other forms of these advisories include winter weather, high wind, flood, tropical cyclone, and fog. Severe weather advisories and alerts are broadcast through the media, including radio, using emergency systems as the Emergency Alert System, which break into regular programming.

Air Traffic

Because the aviation industry is especially sensitive to the weather, accurate weather forecasting is essential considering the fact that a greater number of plane crashes recorded the world over have weather related causes. Just as turbulence and icing are significant in flight hazards, thunderstorms are a major problem for all aircrafts because of severe turbulence due to their updrafts and outflow boundaries, icing due to the heavy precipitation, as well as large hail, strong winds, and lightening, all of which can cause severe damage to an aircraft's in-flight. Volcanic ash is also a significant problem for aviation, as aircrafts can lose engine power with ash clouds. On a day-to-day basis, airliners are routed to take advantage of the jet stream tailwind to improve fuel efficiency. Aircrews are briefed prior to takeoff on the conditions to expect enroute and at their destination. Additionally, airports often change which runway is being used to take advantage of a headwind. This reduces the distance required for takeoff, and to eliminate potential crosswinds.

Marine

Commercial and recreational use of waterways can be limited significantly by wind direction and speed, wave periodicity and heights, tides, and precipitation. These factors can each influence the safety of marine transit. Consequently, a variety of codes have been established to efficiently transmit detailed marine weather forecasts to vessel pilots through radio, for example the MAFOR (Marine forecast).

Agriculture

Farmers rely on weather forecasts to decide what work to do on any particular day. For example, drying hay is only feasible in dry weather. Prolonged periods of dryness can ruin cotton, wheat, and corn crops. While crops can be ruined by drought, their dried remains can be used as a cattle feed substitute in the form of silage. Frosts and freezes play havoc with crops both during the spring and fall. For example, peach tree in full bloom can have their potential peach crop decimated by a spring freeze. Orange groves can suffer significant damage during frosts and freezes, regardless of their timing.

Utility Companies

Electricity and gas companies rely on weather forecasts to anticipate demand, which can be strongly affected by the weather. They use the quantity termed the degree-day to determine how strong of a use there will be for heating (heating degree day) or cooling (cooling degree day). These quantities are based on a daily average temperature of 65 °F (18 °C). Cooler temperatures force heating degree days (one per degree Fahrenheit), while warmer temperatures force cooling degree days. In winter, severe cold weather can cause a surge in demand as people turn up their heating. Similarly, in summer or dry season a surge in demand can be linked with the increased use of air conditioning systems in hot weather. By anticipating a surge in demand, utility companies can produce additional supplies of power or natural gas before the price increases, or in some circumstances, supplies are restricted through the use of brown outs and blackouts.

Private Sector

Increasingly, private companies pay for weather forecasts tailored to their needs so that they can increase their profits or avoid large losses. For example, supermarket chains may change the stocks on their shelves in anticipation of different, consumer spending habits in different weather conditions. Weather forecasts can be used to invest in the commodity market, such as futures in oranges, corn, soybeans and oil. Also, members of the public use knowledge of future weather conditions to determine what to put on, on a daily basis.

Military Applications

Similarly to the private sector, military weather forecasters present weather conditions to the war fighters, community. Military weather forecasters provide pre-flight weather briefs to pilots and provide real time resource protection services for military installations.

Weather Parameters

Dew Point

Dew point is the temperature to which air must be cooled for saturation (100% relative humidity) to occur. The dew point is an important measurement used to predict the formation of dew, frost, and fog. If dew point and temperature are close together in the late afternoon when the air begins to turn colder, fog is likely during the night. Dew forms when the air temperature is equal to or lower than the dew point temperature, and the dew point temperature is above freezing. Frost forms when the air temperature is equal to or lower than the dew point temperature, and the dew point temperature--technically called the frost point temperature--is below freezing. Dew point is also a good indicator of the air's actual water vapor content, unlike relative humidity, which takes the air's temperature into account. High dew point indicates high vapor content; low dew point indicates low vapor content. In addition a high dew point indicates a better chance of rain and severe thunderstorms. You can even use dew point to predict the minimum overnight temperature. Provided no fronts or other weather pattern changes are expected overnight, the afternoons dew point gives you an idea of what minimum temperature to expect overnight, since the air is not likely to get colder than the dew point anytime during the night. This is due to the release of latent heat energy as water vapor changes phase from gas to liquid or gas to solid, the overnight low temperature rarely drops below the dew point temperature. Dew point depression (air temperature minus the dew point temperature) vs. cloud coverage and/or cloud height (In general, when the surface dew point depression is small, skies will be overcast and cloud bases close to the Earth's surface.) Also a dew point of 70 °F or higher makes the conditions uncomfortable for strenuous activities as the body get very little help from evaporative cooling. There is a fine line between being damp and being humid and is determined by the how strenuous the person's activity is at that time.

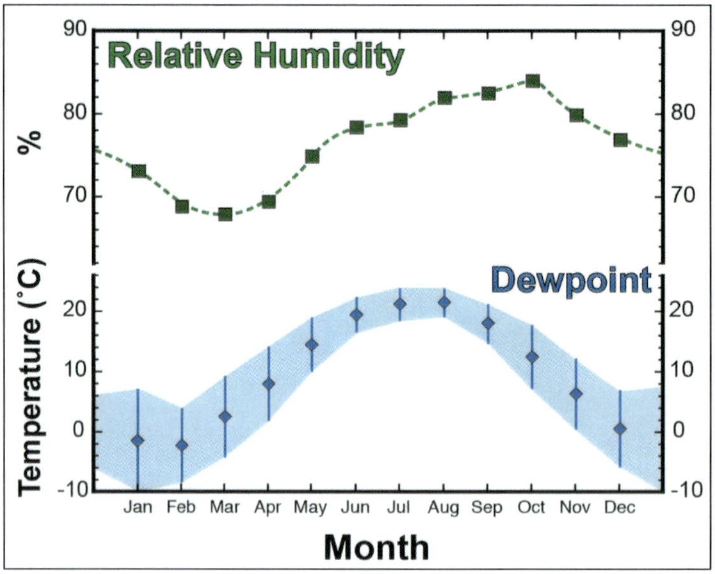

Fog

We have seen heavy fog defined as having a visibility of less than 0.3 miles. But have never seen what defines a moderate or light fog. Is there some standard? This sounds like a pretty simple question, but isn't. First, here's the definition of "fog" from the American Meteorological Society's (AMS) Glossary of Weather and Climate: "A visible aggregate of minute water droplets suspended in the atmosphere near the earth's surface; a cloud in contact with the earth's surface. Fog is responsible for reducing visibility to less than 1 kilometer (5/8 mile)." Aviation weather observations and forecasts, intended for pilots, use "fog" when it reduces visibility to 5/8 statute miles or less. "Mist" is reported and forecast for "foggy conditions" with visibility greater than 5/8 miles. The Weather element allows forecasters to differentiate between F+ (heavy or Dense fog), F (Fog or Moderate Fog), and F- (Light Fog) and the computer worded forecasts pick up on these distinctions. However, there is no national standard for these definitions." He notes that most NWS forecast offices would use the code for heavy fog when the visibility is less than 1/4 mile, just "fog" when the visibility is between 1/4 mile to 1 mile and light fog for visibility of 1 mile or more. "These tend to be set by forecasters based on local climatology for frequency of occurrence."

Barometric Pressure

The air that makes up our atmosphere exerts a pressure on the surface of the earth. This pressure is known as atmospheric pressure. Generally, the more air above an area, the higher the atmospheric pressure. This, in turn, means that atmospheric pressure changes with altitude. For example, atmospheric pressure is greater at sea-level than on a mountaintop. To compensate for this difference in pressure at different elevations, and to facilitate comparison between locations with different altitudes, meteorologists adjust atmospheric pressure so that it reflects what the pressure would be if measured at sea-level. This adjusted pressure is known as barometric pressure.

Barometric pressure changes with local weather conditions, making barometric pressure an important and useful weather forecasting tool. High pressure zones are generally associated with fair weather, while low pressure zones are generally associated with poor weather. For forecasting purposes, the absolute barometric pressure value is generally less important than the change in barometric pressure.

In general, rising pressure indicates improving weather conditions, while falling pressure indicates deteriorating weather conditions.

Air pressure often indicates cloud coverage. Also cloud coverage generally indicates the daily maximum and minimum air temperature range:

- In general, there is a greater temperature range when skies are clear and a smaller temperature range when skies are overcast.
- In general, low air pressure is associated with cloudiness and a chance of precipitation; high pressure is associated with fair weather and clear skies.

Wind is cause by air flowing from higher pressure into lower pressure.

Causes of Pressure Differences

The high and low pressure systems usually develop due to temperature differences. Temperature differences result in pressure differences and this causes the air to move. Thus, the greater the change in the air pressure the greater the possibility of strong winds. When it is windy outside, often there is a low pressure system the air is moving toward. How fast the wind blows will depend on the pressure differences between the high and low pressure systems. Very fast winds often occur near cold fronts, low pressure systems. Wind can also blow faster when it is forced into a narrow space. This can happen between buildings and within mountain passes. Thus the following saying - The higher the high and the lower the low, the faster the wind will flow from the high to the low, and the stronger the winds will blow.

Wind Run

Wind run is measurement of the "amount" of wind passing the station during a given period of time, expressed in either "miles of wind" or "kilometers of wind". WeatherLink calculates wind run by multiplying the average wind speed for each archive record by the archive interval.

For Example:

- Average Wind Speed = 5 mph,
- Archive Interval = 30 minutes (0.5 hours),
- Wind Run = 5 mph x 0.5 hours = 2.5 miles of wind.

A Weather Station uses a 5 minute archive interval. Then the 5 minute archive records are totalled for the 24 hour period from midnight to midnight giving the wind run for the day in miles.

Wind Chill

Wind chill temperature is how cold people and animals feel when outside. Wind chill quantifies the rate of heat loss from exposed skin caused by wind and cold. As the wind increases, it draws heat from the body, driving down skin temperature and eventually the internal body temperature. Therefore, the wind makes it FEEL much colder. If the temperature is 0 °F and the wind is blowing at 15 mph, the wind chill is -19 °F. At this wind chill temperature, exposed skin can freeze in 30 minutes.

The only effect wind chill has on inanimate objects, such as car radiators and water pipes, is to shorten the amount of time for the object to cool. The inanimate object cannot cool below the actual air temperature. For example, if the temperature outside is -5 °F and the wind chill temperature is -31 °F, then your car's radiator will not drop lower than -5 °F.

Precipitation Rate

The precipitation rate is calculated by measuring the time interval between each measured precipitation increment of .01". When there is precipitation within the archive period, the highest measured value is reported. When precipitation ends, the rain rate will slowly decay based on the elapse time since the last measured precipitation. If no precipitation is measured for 15 minutes the precipitation rate is reset to zero.

Evaporation

Evaporation is the estimated amount of water vapor returned to the air based on measured temperature, relative humidity, average wind speed, and solar radiation.

Wet Bulb Temperature

The wet bulb temperature is a measure of the amount of moisture, in the form of invisible water vapor contained in the air. As the name implies it is measured by a standard thermometer whose bulb is covered by a muslin sleeve that has been moistened by pure water. This is the temperature air can be cooled to by evaporating water into it and therefore the wet bulb temperature provides a good estimate of how far temperatures will fall if it starts raining or snowing.

The principle of the wet bulb thermometer is as follows; water evaporates from the muslin cover passing into the air in the form of invisible water vapor. In so doing it absorbs heat from the thermometer bulb. The thermometer therefore indicates a lower temperature than that of the dry bulb thermometer. The difference between the readings of the dry and the wet thermometers is called the depression of the wet bulb.

If the air contains nearly all the moisture it can possibly hold, evaporation from the muslin will be slight and the depression of the wet bulb will be small. However, if the air is very dry, containing little moisture, evaporation will be quite rapid and the depression of the wet bulb will be quite large. In hot dry desert climates depressions of over 25 °C have been observed, but at sea the depression is seldom more than 5 °C. If the air contains all the moisture it can possibly hold, there is no evaporation from the muslin, and the dry and wet bulb thermometers will read the same. When this condition exists the air is said to be saturated.

Provided that the wet bulb is adequately moistened and given proper ventilation, its reading will always be equal to or less than that of the dry bulb when the air temperature is above freezing. Under certain conditions when the air temperature is below freezing and there is ice on the wet bulb, its reading may be slightly higher than the dry bulb. This is called a negative depression and usually occurs with fog or precipitation. Negative depressions are rare.

Soil Temperature

Soil temperature is reported to the nearest degree Fahrenheit and is measured at a depth of 5 inches and 12 inches depth beneath a grassy surface.

Forecasts of Weather Parameters

Maximum Temperature Forecast

The forecast of maximum temperature is based on sunshine during the day. Cold and warm advections have pronounced effect on the temperature conditions. Air temperature is very important from agriculture point of view. Temperature provides working conditions for the growth of crop plants. For the normal growth, the temperature should be optimum.

There are extreme temperatures beyond this optimum. These extremes are called maximum air temperature and minimum air temperature. The forecast of these extreme temperatures is always helpful for taking precautionary measures to save the crops being subjected to highest and lowest temperature.

Minimum Temperature Forecast

The forecast of minimum temperature is based on the cloudy conditions during night. Cold and warm advections have pronounced effect on the night time temperature. During winter season minimum temperature remains close to freezing point. In order to save the crops from low temperature injury, the forecast of minimum temperature is very important.

Predicted minimum temperature should not be less than expected dew point temperature. During summer season, temperature rises and it becomes maximum during the months of May and June. Temperature increases abruptly when warm advection takes place. As a result, heat wave conditions are intensified which are harmful for the crop plants.

Similarly during winter season, temperature decreases due to cold advection. It leads to cold wave conditions and frosty nights. Such types of weather conditions are quite prevalent in north-west India. The frosty nights have adverse effect on the growth of crop plants. The crop plants, which are susceptible to low temperature conditions, suffer maximum damage when air temperature falls below freezing level.

Wind Forecast

Wind is generated by the movement of air-mass due to the horizontal pressure gradient set up between the two stations. The forecast of wind is made based on the spacing of the isobars. Closer the isobars, higher the strength of the winds and vice-versa. Generally strong surface winds are associated with low pressure areas or cyclonic circulation.

If the low pressure area is intensified and becomes cyclone, then the strength of the winds becomes very strong and at times may exceed even 64 kmph. On the other hand, light winds are associated with high pressure area or anticyclone circulation.

Wind forecast depends upon the pressure distribution on the surface chart. Closer the isobars, stronger would be the wind speed. The maximum wind speed at the surface cannot exceed the highest wind speed at 850 mb. Wind forecast is important for spraying chemicals and fertilizer application.

Harvesting efficiency increases under calm conditions. In view of the anticipated windy condition, irrigation is avoided; otherwise lodging of the crop plants takes place. Generally stable weather conditions prevail in the morning and in the evening hours. Therefore, calm conditions associated with low temperature are favourable for spraying insecticides, fungicides and herbicides.

Cloud Forecast/Precipitation

The information on these weather parameters is useful for farmers in taking decision on agricultural operations like spraying, irrigation and harvesting. Sunshine is also required for agricultural operations like threshing and drying the agricultural produce.

The forecast of the cloud cover depends upon the synoptic situation. If the area is under the influence of low pressure associated with upper air cyclonic circulation or upper air trough, then cloudiness or precipitation may be expected depending upon the available water vapours.

Rainfall is described in the forecast as isolated, where less than one-third of the recording stations are expected to receive rain. It is described scattered, when one-third to two-third stations are expected to record rain and wide spread, when more than two third stations are expected to experience rainfall.

Conditions for Clear Weather

- There must be anticyclone i.e. dry and cold winds must blow in clockwise direction.
- There must be negative vertical motion. Descending air mass becomes hot due to adiabatic warming.
- Relative humidity should be less than 50 per cent.

Dew Forecast

Dew appears in the morning hours on vegetation near the earth's surface, when the air temperature falls to dew point temperature. The condensation of water vapours from the air upon the vegetation surface cooled by radiation losses is called 'Dew'. The amount of dew is very small as compared to precipitation.

There are large variations in the quantity of dew deposited on the surface. Dew is a form of precipitation. It is very important for agricultural plants in those areas where scanty rainfall occurs. In the desert areas, dew provides moisture to the vegetation for its survival.

Crops grown under rained conditions are greatly benefited from the moisture provided by dew in the morning hours. But dew is favourable for the incidence of plant diseases in the crops grown under irrigated conditions.

Therefore, dew forecast can be made in different categories. According to the pictorial guide devised by Dudevani, dew can be divided into six categories. These are very light, light, moderate, heavy, and very heavy and wet.

Conditions Favourable for Dew

1. Clear sky and little surface moisture.
2. Light winds or wind speed should be less than 8 kmph.
3. Rapid outgoing radiational cooling at night.
4. Presence of high pressure area or anticyclone winds.
5. Absence of positive vertical motion.
6. Absence of turbulence.

If these conditions are fulfilled then dew is predicted.

Conditions not Favourable for Dew

1. Cloudy conditions are favourable for the reflection of long wave radiation back to the earth.
2. Presence of low pressure area or cyclonic circulation.
3. Strong surface winds i.e. wind speed should be greater than 16 kmph.
4. Presence of negative vertical motion. Descending air does not allow the air temperature to reach dew point.

Evaporation

The rate of evaporation depends upon the prevailing weather conditions and the anticipated weather conditions. Forecasting of evaporation is required to apply irrigation to the crops. Application of irrigation depends upon the type of soil and the crop grown. It is simply based on the amount of accumulated evaporation since the date of application of irrigation. The loss of water depends upon the weather conditions.

The atmospheric demand increases under hot and dry weather conditions and decreases under humid weather conditions. The prediction of evaporation is helpful in planning irrigation in advance so that water can be used judiciously.

Frost Forecast

Frost is the form of precipitation. Like dew, frost also plays an important role in agriculture. The formation of frost depends upon the dew point. When the dew point decreases below freezing temperature, then the water vapours directly change to ice crystals. This process is called sublimation. The layer of the ice crystals is called frost. The frost causes low temperature injury to the crop plants.

The vegetable plants are sensitive to frost injury. Frost generally occurs during winter season.

Conditions favourable for the formation of frost:

1. Sky should be clear;
2. There must be a high pressure area or anticyclone wind;
3. Calm conditions;
4. Excessive radiational cooling;
5. Low water content in the air;
6. Absence of turbulence;
7. Absence of positive vertical motion.

These type of conditions are generally found when the dry weather continues for a few days. Therefore, the forecast of frost is very important. The farmers can adopt protection measures, if the forecast of frost is made well in advance. The prevention of frost and protection of plants from damaging frost can be achieved by using various methods.

Reference

- Weather, shortcontent: scied.ucar.edu, Retrieved 6 August, 2019
- Condensation, encyclopedia: nationalgeographic.org, Retrieved 5 July, 2019
- Forecasts-of-weather-parameters, climatology: yourarticlelibrary.com, Retrieved 4 June, 2019
- Wxterms, alink: glenallenweather.com, Retrieved 5 May, 2019

5

Instruments and Equipments used in Meteorology

There are a number of instruments and equipments which are used in meteorology such as anemometer, barometer, ceilometer, hygrometer, weather balloon, weather radar, pyranometer and rain gauge. The topics elaborated in this chapter will help in gaining a better perspective about these instruments and equipment used in meteorology.

Anemometer

An anemometer is a device for measuring wind speed, and is a common weather station instrument. The first known description of an anemometer was given by Leon Battista Alberti around 1450. They are also very easy to make as a project.

Anemometers can be divided into two classes: those that measure the wind's speed, and those that measure the wind's pressure; but as there is a close connection between the pressure and the speed, an anemometer designed for one will give information about both.

A hemispherical cup anemometer of the type invented in 1846 by John Thomas Romney Robinson.

Velocity Anemometers

Cup Anemometers

A simple type of anemometer, invented by Dr. John Thomas Romney Robinson, of Armagh Observatory. It consisted of four hemispherical cups each mounted on one end of four horizontal arms, which in turn were mounted at equal angles to each other on a vertical shaft. The air flow past the cups in any horizontal direction turned the cups in a manner that was proportional to the wind speed. Therefore, counting the turns of the cups over a set time period produced the average wind speed for a wide range of speeds. On an anemometer with four cups it is easy to see that since the cups are arranged symmetrically on the end of the arms, the wind always has the hollow of one cup presented to it and is blowing on the back of the cup on the opposite end of the cross.

Cup-type anemometer with vertical axis, a sensor on a remote meteorological station deployed on Skagit Bay, Washington July–August, 2009.

When Robinson first designed his anemometer, he asserted that the cups moved one-third of the speed of the wind, unaffected by the cup size or arm length. This was apparently confirmed by some early independent experiments, but it was incorrect. Instead, the ratio of the speed of the wind and that of the cups, the anemometer factor, depends on the dimensions of the cups and arms, and may have a value between two and a little over three. Every experiment involving an anemometer had to be repeated.

The three cup anemometer developed by the Canadian John Patterson in 1926 and subsequent cup improvements by Brevoort and Joiner of the USA in 1935 led to a cupwheel design which was linear and had an error of less than 3% up to 60 mph (97 km/h). Patterson found that each cup produced maximum torque when it was at 45 degrees to the wind flow. The three cup anemometer also had a more constant torque and responded more quickly to gusts than the four cup anemometer.

The three cup anemometer was further modified by the Australian Derek Weston in 1991 to measure both wind direction and wind speed. Weston added a tag to one cup, which causes the cupwheel speed to increase and decrease as the tag moves alternately with and against the wind. Wind direction is calculated from these cyclical changes in cupwheel speed, while wind speed is as usual determined from the average cupwheel speed. Three cup anemometers are currently used as the industry standard for wind resource assessment studies.

Windmill Anemometers

The other forms of mechanical velocity anemometer may be described as belonging to the windmill type or propeller anemometer. In the Robinson anemometer the axis of rotation is vertical, but with this subdivision the axis of rotation must be parallel to the direction of the wind and therefore horizontal. Furthermore, since the wind varies in direction and the axis has to follow its changes, a wind vane or some other contrivance to fulfil the same purpose must be employed. An aerovane combines a propeller and a tail on the same axis to obtain accurate and precise wind speed and direction measurements from the same instrument. In cases where the direction of the air motion is always the same, as in the ventilating shafts of mines and buildings for instance, wind vanes, known as air meters are employed, and give most satisfactory results.

A windmill style of anemometer.

Hot-wire Anemometers

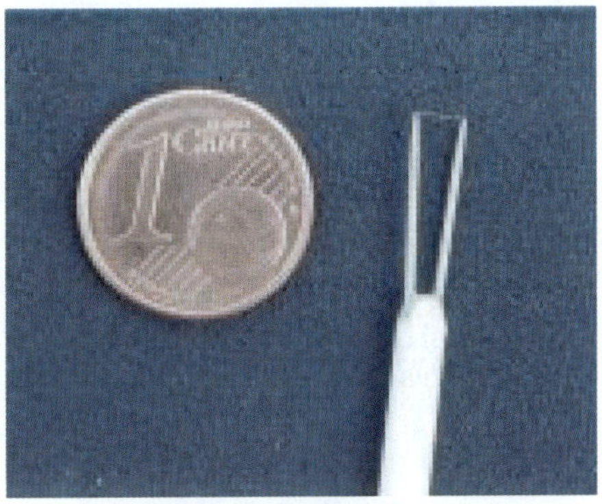

Hot-wire sensor.

Hot wire anemometers use a very fine wire (on the order of several micrometres) electrically heated up to some temperature above the ambient. Air flowing past the wire has a cooling effect on the wire. As the electrical resistance of most metals is dependent upon the temperature of the metal (tungsten is a popular choice for hot-wires), a relationship can be obtained between the resistance of the wire and the flow speed.

Several ways of implementing this exist, and hot-wire devices can be further classified as CCA (Constant-Current Anemometer), CVA (Constant-Voltage Anemometer) and CTA (Constant-Temperature Anemometer). The voltage output from these anemometers is thus the result of some sort of circuit within the device trying to maintain the specific variable (current, voltage or temperature) constant.

Additionally, PWM (pulse-width modulation) anemometers are also used, wherein the velocity is inferred by the time length of a repeating pulse of current that brings the wire up to a specified resistance and then stops until a threshold "floor" is reached, at which time the pulse is sent again.

Hot-wire anemometers, while extremely delicate, have extremely high frequency-response and fine spatial resolution compared to other measurement methods, and as such are almost universally employed for the detailed study of turbulent flows, or any flow in which rapid velocity fluctuations are of interest.

Laser Doppler Anemometers

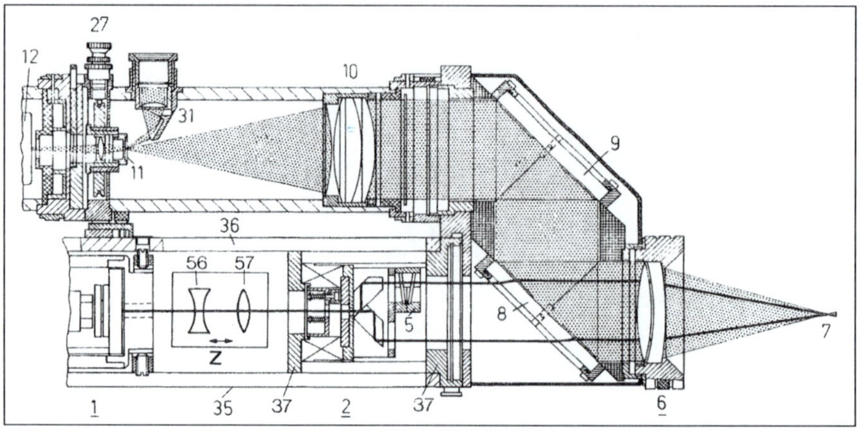

Drawing of a laser anemometer. The laser is emitted (1) through the front lens (6) of the anemometer and is backscattered off the air molecules (7). The backscattered radiation (dots) re-enter the device and are reflected and directed into a detector (12).

Laser Doppler anemometers use a beam of light from a laser that is divided into two beams, with one propagated out of the anemometer. Particulates (or deliberately introduced seed material) flowing along with air molecules near where the beam exits reflect, or backscatter, the light back into a detector, where it is measured relative to the original laser beam. When the particles are in great motion, they produce a Doppler shift for measuring wind speed in the laser light, which is used to calculate the speed of the particles, and therefore the air around the anemometer.

Sonic Anemometers

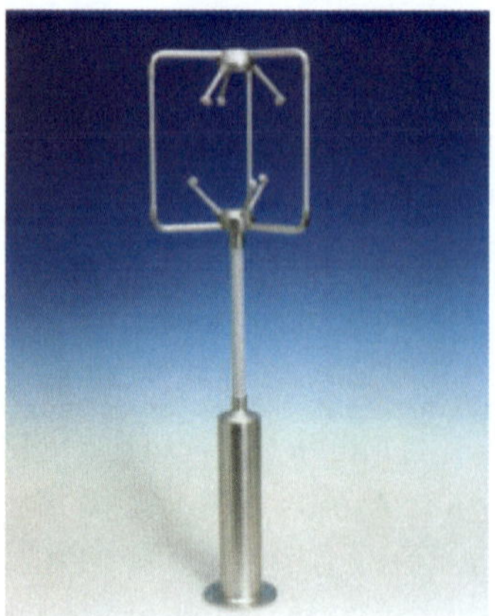

3D ultrasonic anemometer.

Sonic anemometers, first developed in the 1970s, use ultrasonic sound waves to measure wind velocity. They measure wind speed based on the time of flight of sonic pulses between pairs of transducers. Measurements from pairs of transducers can be combined to yield a measurement of velocity in 1-, 2-, or 3-dimensional flow. The spatial resolution is given by the path length between transducers, which is typically 10 to 20 cm. Sonic anemometers can take measurements with very fine temporal resolution, 20 Hz or better, which makes them well suited for turbulence measurements. The lack of moving parts makes them appropriate for long term use in exposed automated weather stations and weather buoys where the accuracy and reliability of traditional cup-and-vane anemometers is adversely affected by salty air or large amounts of dust. Their main disadvantage is the distortion of the flow itself by the structure supporting the transducers, which requires a correction based upon wind tunnel measurements to minimize the effect. An international standard for this process, ISO 16622 Meteorology—Sonic anemometers/thermometers—Acceptance test methods for mean wind measurements is in general circulation. Another disadvantage is lower accuracy due to precipitation, where rain drops may vary the speed of sound.

Since the speed of sound varies with temperature, and is virtually stable with pressure change, sonic anomometers are also used as thermometers.

Two-dimensional (wind speed and wind direction) sonic anemometers are used in applications such as weather stations, ship navigation, wind turbines, aviation and weather buoys.

Ping-pong Ball Anemometers

A common anemometer for basic use is constructed from a ping-pong ball attached to a string. When the wind blows horizontally, it presses on and moves the ball; because ping-pong balls are very lightweight, they move easily in light winds. Measuring the angle between the string-ball apparatus and the line normal to the ground gives an estimate of the wind speed.

This type of anemometer is mostly used for middle-school level instruction which most students make themselves, but a similar device was also flown on Phoenix Mars Lander.

Pressure Anemometers

The first designs of anemometers which measure the pressure were divided into plate and tube classes.

Plate Anemometers

These are the earliest anemometers and are simply a flat plate suspended from the top so that the wind deflects the plate. In 1450, the Italian art architect Leon Battista Alberti invented the first mechanical anemometer; in 1664 it was re-invented by Robert Hooke (who is often mistakenly considered the inventor of the first anemometer). Later versions of this form consisted of a flat plate, either square or circular, which is kept normal to the wind by a wind vane. The pressure of the wind on its face is balanced by a spring. The compression of the spring determines the actual force which the wind is exerting on the plate, and this is either read off on a suitable gauge, or on a recorder. Instruments of this kind do not respond to light winds, are inaccurate for high wind readings, and are slow at responding to variable winds. Plate anemometers have been used to trigger high wind alarms on bridges.

Tube Anemometers

James Lind's anemometer of 1775 consisted simply of a glass U tube containing liquid, a manometer, with one end bent in a horizontal direction to face the wind and the other vertical end remains parallel to the wind flow. Though the Lind was not the first it was the most practical and best known anemometer of this type. If the wind blows into the mouth of a tube it causes an increase of pressure on one side of the manometer. The wind over the open end of a vertical tube causes little change in pressure on the other side of the manometer. The resulting liquid change in the U tube is an indication of the wind speed. A small departure from the true direction of the wind causes large variations in the magnitude.

Helicoid propeller anemometer incorporating a wind vane for orientation.

The highly successful metal pressure tube anemometer of William Henry Dines in 1892 utilized the same pressure difference between the open mouth of a straight tube facing the wind and a ring of small holes in a vertical tube which is closed at the upper end. Both are mounted at the same height. The pressure differences on which the action depends are very small, and special means are required to register them. The recorder consists of a float in a sealed chamber partially filled with water. The pipe from the straight tube is connected to the top of the sealed chamber and the pipe from the small tubes is directed into the bottom inside the float. Since the pressure difference determines the vertical position of the float this is a measure of the wind speed.

The great advantage of the tube anemometer lies in the fact that the exposed part can be mounted on a high pole, and requires no oiling or attention for years; and the registering part can be placed in any convenient position. Two connecting tubes are required. It might appear at first sight as though one connection would serve, but the differences in pressure on which these instruments depend are so minute, that the pressure of the air in the room where the recording part is placed has to be considered. Thus if the instrument depends on the pressure or suction effect alone, and this pressure or suction is measured against the air pressure in an ordinary room, in which the doors and windows are carefully closed and a newspaper is then burnt up the chimney, an effect may be produced equal to a wind of 10 mi/h (16 km/h); and the opening of a window in rough weather, or the opening of a door, may entirely alter the registration.

While the Dines anemometer had an error of only 1% at 10 mph (16 km/h) it did not respond very well to low winds due to the poor response of the flat plate vane required to turn the head into the wind. In 1918 an aerodynamic vane with eight times the torque of the flat plate overcame this problem.

Effect of Density on Measurements

In the tube anemometer the pressure is measured, although the scale is usually graduated as a velocity scale. In cases where the density of the air is significantly different from the calibration value (as on a high mountain, or with an exceptionally low barometer) an allowance must be made. Approximately 1½% should be added to the velocity recorded by a tube anemometer for each 1000 ft (5% for each kilometer) above sea-level.

Barometer

A barometer is a scientific instrument used to measure atmospheric pressure, also called barometric pressure. The atmosphere is the layers of air wrapped around the Earth. That air has a weight and presses against everything it touches as gravity pulls it to Earth. Barometers measure this pressure.

Atmospheric pressure is an indicator of weather. Changes in the atmosphere, including changes in air pressure, affect the weather. Meteorologists use barometers to predict short-term changes in the weather.

A rapid drop in atmospheric pressure means that a low-pressure system is arriving. Low pressure means that there isn't enough force, or pressure, to push clouds or storms away. Low-pressure systems are associated with cloudy, rainy, or windy weather. A rapid increase in atmospheric pressure pushes that cloudy and rainy weather out, clearing the skies and bringing in cool, dry air.

A barometer measures atmospheric pressure in units of measurement called atmospheres or bars. An atmosphere (atm) is a unit of measurement equal to the average air pressure at sea level at a temperature of 15 degrees Celsius (59 degrees Fahrenheit).

The number of atmospheres drops as altitude increases because the density of air is lower and exerts less pressure. As altitude decreases, the density of air increases, as does the number of atmospheres. Barometers have to be adjusted for changes in altitude in order to make accurate atmospheric pressure readings.

Types of Barometers

Mercury Barometer

The mercury barometer is the oldest type of barometer, invented by the Italian physicist Evangelista Torricelli in 1643. Torricelli conducted his first barometric experiments using a tube of water. Water is relatively light in weight, so a very tall tube with a large amount of water had to be used in order to compensate for the heavier weight of atmospheric pressure.

Torricelli's water barometer was more than 10 meters (35 feet) in height, which rose above the roof of his home. This odd device caused suspicion among Torricelli's neighbors, who thought he was involved in witchcraft. In order to keep his experiments more secretive, Torricelli deduced that he could create a much smaller barometer using mercury, a silvery liquid that weighs 14 times as much as water.

A mercury barometer has a glass tube that is closed at the top and opens at the bottom. At the bottom of the tube is a pool of mercury. The mercury sits in a circular, shallow dish surrounding the tube. The mercury in the tube will adjust itself to match the atmospheric pressure above the dish. As the pressure increases, it forces the mercury up the tube. The tube is marked with a series of measurements that track the number of atmospheres or bars. Observers can tell what the air pressure is by looking at where the mercury stops in the barometer.

Aneroid Barometer

In 1844, the French scientist Lucien Vidi invented the aneroid barometer. An aneroid barometer has a sealed metal chamber that expands and contracts, depending on the atmospheric pressure around it. Mechanical tools measure how much the chamber expands or contracts. These measurements are aligned with atmospheres or bars.

The aneroid barometer has a circular display that indicates the present number of atmospheres, much like a clock. One hand moves clockwise or counter clockwise to point to the current number of atmospheres. The terms stormy, rain, change, fair, and dry are often written above the numbers on the dial face to make it easier for people to interpret the weather. Aneroid barometers slowly replaced mercury barometers because they were easier to use, cheaper to buy, and easier to transport since they had no liquid that could spill.

Some aneroid barometers use a mechanical tool to track the changes in atmospheric pressure over a period of time. These aneroid barometers are called barographs. Barographs are barometers connected to needles that make marks on a roll of adjacent graph paper. The barograph records the number of atmospheres on the vertical axis and units of time on the horizontal. A barograph's tracking tool will rotate, usually once every day, week, or month. The spikes in the graph show when air pressure was high or low, and how long those pressure systems lasted. A severe storm, for instance, would appear as a deep, wide dip on a barograph.

Digital Barometers

Today's digital barometers measure and display complex atmospheric data more accurately and quickly than ever before. Many digital barometers display current barometric readings and previous 1-, 3-, 6-, and 12-hour readings in a bar chart format, much like a barograph. They also account for other atmospheric readings such as wind and humidity to make accurate weather forecasts. This data is archived and stored on the barometer and can also be downloaded onto a computer for further analysis. Digital barometers are used by meteorologists and other scientists who want up-to-date atmospheric readings when conducting experiments in the lab or out in the field.

The digital barometer is now an important tool in many of today's smartphones. This type of digital barometer uses atmospheric pressure data to make accurate elevation readings. These readings help the smartphone's GPS receiver pinpoint a location more accurately, greatly improving navigation.

Developers and researchers are also using the smartphone's crowdsourcing capabilities to make more accurate weather forecasts. Apps like Pressure Net automatically collect barometric measurements from each of its users, creating a vast network of atmospheric data. This data network makes it easier and faster to map out storms as they develop, especially in areas with few weather stations.

Ceilometer

Ceilometer is device for measuring the height of cloud bases and overall cloud thickness. One important use of the ceilometer is to determine cloud ceilings at airports. The device works day or night by shining an intense beam of light (often produced by an infrared or ultraviolet transmitter or a laser), modulated at an audio frequency, at overhead clouds. Reflections of this light from the base of the clouds are detected by a photocell in the receiver of the ceilometer. There are two basic types of ceilometers: the scanning receiver and the rotating transmitter.

The scanning-receiver ceilometer has its separate light transmitter fixed to direct its beam vertically. The receiver is stationed a known distance away. The parabolic collector of the receiver continuously scans up and down the vertical beam, searching for the point where the light intersects a cloud base. When a reflection is detected, the ceilometer measures the vertical angle to the spot; a simple trigonometric calculation then yields the height of the cloud ceiling. Many modern scanning-receiver ceilometers use a laser pulse to identify the height of a cloud's base and top and various points in between to create a vertical profile of the cloud.

The rotating-transmitter ceilometer has its separate receiver fixed to direct reflections only from directly overhead while the transmitter sweeps the sky. When the modulated beam intersects a cloud base directly over the receiver, light is reflected downward and detected.

Optical Drum Ceilometer

An optical drum ceilometer uses triangulation to determine the height of a spot of light projected onto the base of the cloud. It consists essentially of a rotating projector, a detector, and a recorder. The projector emits an intense beam of light above into the sky at an angle that varies with the rotation. The detector, which is located at a fixed distance from the projector, uses a photoelectric cell pointing vertically. When it detects the projected light return from the cloud base, the instrument notes the angle and the calculation gives the height of clouds.

Laser Ceilometer

Laser ceilometers consist of a vertically pointing laser and a receiver in the same location. It determines the height by measuring the time (delta t) required for a pulse of light to be scattered back from aerosols within the atmosphere.

Generally, the sizes of the particles in question are similar in size to the wavelength of the laser. This situation leads to MIE theory.

For cloud base determination purpose, due to the ceilometers ability to pick up any particle in the air (dust, precipitation, smoke, etc.), it will give occasional false readings. As an example, depending on the threshold used, falling Diamond dust (Ice Crystals) may cause the ceilometer to report a cloud height of zero, even though the sky is clear.

Using these last properties, ceilometers will have other uses. Since the instrument will note any returns, it is possible to locate any faint layer where it occurs, additionally to the cloud's base, by looking at the whole pattern of returned energy. Furthermore, the rate at which diffusion happens can be noted by the diminishing part returned to the ceilometer in clear air, giving the coefficient of extinction of the light signal. Using these data could give the vertical visibility and the possible concentration of air pollutants. This has been developed in research and could be applied for operational purpose.

Hygrometer

A hygrometer is an instrument used to measure relative humidity. It is also called a psychrometer. In order to determine relative humidity, the hygrometer makes use of wet and dry bulb thermometers. The differences in temperatures between the two measurements are an indicator of evaporation, therefore also relative humidity. Older hygrometers were used with the aid of a chart, which was used for the interpretation of readings, but digital hygrometers are more commonly used now. In order to maintain workplace air quality, relative humidity should be 20 - 60%.

Health and safety professionals are required to monitor indoor air quality, record the results and make recommendations if air quality falls outside acceptable parameters. The amount of moisture in the air affects respiratory health, as well as physical comfort. The measurement of relative humidity is therefore an important element in the maintenance of indoor air quality. An OSHA approved hygrometer should be used, since the accuracy of commercial hygrometers is variable. Humidity measurements should be recorded together with temperature readings, and workplace ventilation systems should be tested regularly.

Types of Hygrometers

Hygrometers are used to measure moisture content in the surrounding environment. This instrument needs to measure additional variables including temperature, pressure, mass, and the transfer of an electrical charge in order to provide an accurate reading on humidity content in a given sample. An accurate measure of humidity also relies on calibration of the hygrometer. There are three types of hygrometer instruments, all engineered in a slightly different way, but aim to produce a clear measurement of humidity content.

Capacitive Humidity Detector

A capacity humidity sensing device is designed with a polymer film sandwiched between metal conductive electrodes. These sensors are widely applied in industrial settings and as weather telemetry instruments. Figure illustrates the functional structure to a typical capacitive humidity sensor.

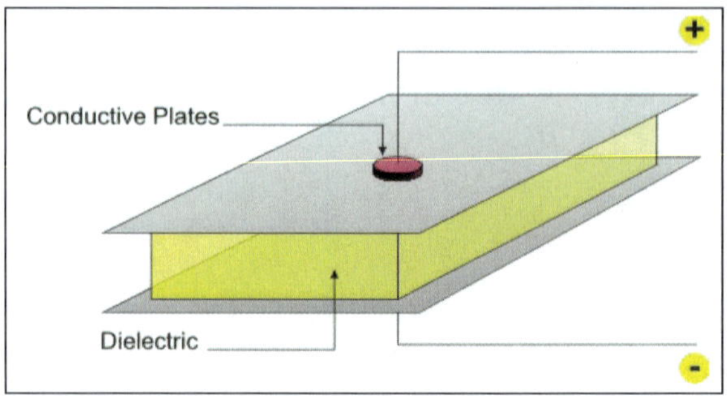

Capacitor Structure.

The capacitive relative humidity (RH) sensor is formed by merging two conductive plates that sandwich an insulator referred to as a dielectric. In this instance, a capacitance is calculated by measuring the area of both plates, the length of space and the dielectric coefficient. The purpose of the insulator is to absorb any water vapour diffusing through into the RH sensor. A high shift in the electrical charge from an average equilibrium position is directly proportional to the relative humidity content in a sample taken from the environment. Typically, a divergence in the capacitance from 0.2 to 0.5 pF is observable for a one per cent change in RH. So, bearing in mind that the electrical capacitance will be higher when the dielectric collects more water, the RH of the surrounding environment will therefore be proportional to the dielectric coefficient. Such capacitive humidity sensors are engineered to work in an extreme temperature range and are resistant to chemical vapours.

Advantages

- Linear voltage output.
- Works well under a wide temperature range and maintains condensation tolerance.
- Ideal for long-term application.

Disadvantages

- The distance between the humidity sensor and the signaling circuit is limited and so this restricts its application over large open areas.

Hair-tension Humidity Sensors

Compared to the RH sensor, a more organic form of a typical humidity sensor is that of the hair tension hygrometer. This type of hygrometer uses animal hair mounted in clusters as sensing elements. Keeping the hair under tension increases the sensitivity of the hair to change in humidity, temperature, thus a direct change in the length of the hair can be measured by using a dial or scale. The tips of each hair are bundled together and held in position using a screw. The lower end of each hair is aligned and attached to an adjustable arm which keeps each hair under tension. The screw at the top end of this hygrometer is also adjustable for shifts.

Each hair to this device is approximately 200 to 300 mm in length. Any change in atmospheric temperature and humidity will change the length of each hair connected to an array of levers wired to a recording device. In response to change in humidity, hair expands or contracts which displaces the movement of a hook. The movement of each hair is magnified which provides an accurate and linear measurement of relative humidity (RH) over 0 to 100%.

Advantages

- Application of this device has been popular amongst meteorological research based on the simplicity of its use.

Disadvantages

- Rapid changes in humidity magnify the errors recorded by the dry-hair hygrometer.

- Increased sensitivity to rapid changes in humidity limits its use in extreme environments where sudden shifts in the level of humidity are common.
- The hair samples can become damaged due to continuous tension.

Electronic Humidity Sensor

Modern-day humidity sensors are based on microprocessor-based circuitry. For example, the Dew point humidity sensor is stable and can be used over a long period of time. Dew point is defined as the temperature state that allows air moisture to saturate as vapour. A working principle for this sensor involves a mirror surface and optoelectronics to measure the amount of water vapour on the surface of the mirror. The temperature to this mirror is electronically monitored to help measure the dew point. A monolithic circuit is connected to the dew point sensor and provides an electrical wave proportional to RH which helps calculate vapour pressure.

Advantages:

- Durable;
- Resistant to chemical vapours;
- Can be applied in extreme temperature environments (e.g., 575 °F);
- You can make a direct comparison of a dew point measurement and the relative temperature of material.

Disadvantages:

- Gases with thermal properties will obscure the overall reading on humidity content measurements.

There are a number of leading sensor manufacturers have introduced a range of humidity sensor lines including Honeywell and Sensirion.

Applications for Humidity Sensors

Humidity sensors have a wide range of applications, such as:

- Food processing industry (e.g., incubators for eggs);
- Social health settings, such as saunas;
- Coating industry where the quality of paint needs to be controlled due to it being sensitive to dew point conditions;
- Meteorological research;
- Environmental monitoring (e.g., to help restore environmentally sensitive rainforests);
- Automotive on-board devices (i.e., windshield defoggers);
- Medical (e.g., ventilators and incubators);
- Consumer electronics;
- Building technology (e.g., cavity wall insulation).

Pyranometer

A pyranometer is a type of actinometer that can measure solar irradiance in the desired location and solar radiation flux density. The solar radiation spectrum extends approximately between 300 and 2800 nm. The pyranometer only requires a flat spectral sensitivity to help cover this spectrum.

The World Meteorological Organization has adopted this instrument which is standardized with respect to the ISO 9060 standard. Pyranometers are calibrated based on the World Radiometric Reference, which is maintained by World Radiation Center, Davos, Switzerland.

Design of Pyranometers

The following are the two main components of pyranometers:

- Thermopile: A sensor consisting of thermocouples connected in series and provided with a black coating for absorbing all solar radiation. It exhibits a near-perfect cosine response and a flat spectrum that covers 300 to 50000 nm. It is capable of producing a potential that is relative to the temperature gradient.
- Glass dome: This dome restricts the spectral response from 300 to 2800 nm from a field of view of 180 degrees. This hemispherical glass dome also shields the thermopile from wind, rain and convection.
- Occulting disc: It is used for measuring the diffuse radiation and blocking beam radiation from the surface.

Working Principle

A pyranometer is operated based on the measurement of temperature difference between a clear surface and a dark surface.

The black coating on the thermopile sensor absorbs the solar radiation, while the clear surface reflects it, and hence less heat is absorbed.

The thermopile is used to measure this temperature difference. The potential difference created in the thermopile owing to the temperature gradient between the two surfaces is used for measuring the amount of solar radiation.

However, the voltage produced by the thermopile can be measured using a potentiometer. Radiation data needs to be integrated by means of an electronic integrator or planimetry.

Advantages

The key benefits of pyranometers are listed below:

- Very small temperature coefficient.
- Calibrated to ISO standards.
- More accurate measurements of performance index and performance ratio.

- Longer response time than a photovoltaic cell.
- Integrated measurement of the total available short-wave solar energy under all conditions.

Applications

The following are the major applications of pyranometers:

- Predicting insulation requirements for building structures.
- Establishment of greenhouse locations.
- Designing photovoltaic systems.
- Meteorological and climatological studies.
- Measurement of solar intensity data.

Rain Gauge

A rain gauge is a meteorological instrument to measure the precipitating rain in a given amount of time per unit area. The instrument consists of a collection container which is placed in an open area. The precipitation is measured in terms of the height of the precipitated water accumulated in the container per given time and is expressed in millimetres. Since the same amount of rain precipitation is assumed to be occurring around the container, the area of collection is not a factor. However, it should not be too small; neither should it be too large. Due to spatial uniformity of rainfall, 1 mm of measured precipitation is the equivalent of 1 L of precipitated rain water volume per metre squared.

There are two types of rain gauges, non-recording type and recording type. Rain gauge is a type of instrument used by meteorologists and hydrologists to measure rainfall rate in a certain period of time. Rain gauges are also known as udometer, pluviometer and ombrometer.

Types of Rain Gauges

Non-recording Type Rain Gauge

Example: Symons Rain Gauge. Non-recording type rain gauge is most common type of rain gauge used by meteorological department. It consists of a cylindrical vessel 127mm in diameter with a base enlarged to 210mm diameter.

At its top section, funnel is provided with circular brass rim which is 127mm exactly so that it can fit into vessel well. This funnel shank is inserted in the neck of a receiving bottle which is 75 to 100mm high from the base section and thinner than the cylinder, placed into it to receive rainfall.

A Receiving bottle has capacity of 100mm and during heavy rainfall, amount of rain is frequently exceeded, so the reading should be measured 3 to 4 times in a day. Water contained in this receiving bottle is measured by a graduated measuring glass with accuracy up to 0. 1mm. For uniformity the rainfall is measured every day at 8:30am IST and is recorded as rainfall of the day.

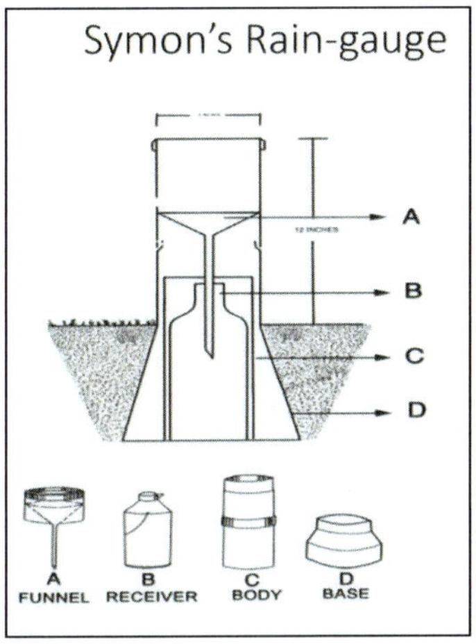

Proper care, maintenance and inspection of rain gauge especially during dry weather are necessary to keep the instrument free form dust and dirt, so that the readings are accurate.

Recording Type Rain Gauges

There are three types of recording rain gauges:

- Weighing bucket type,
- Tipping bucket type,
- Floating or natural syphon type rain gauge.

Weighing Bucket Type Rain Gauge

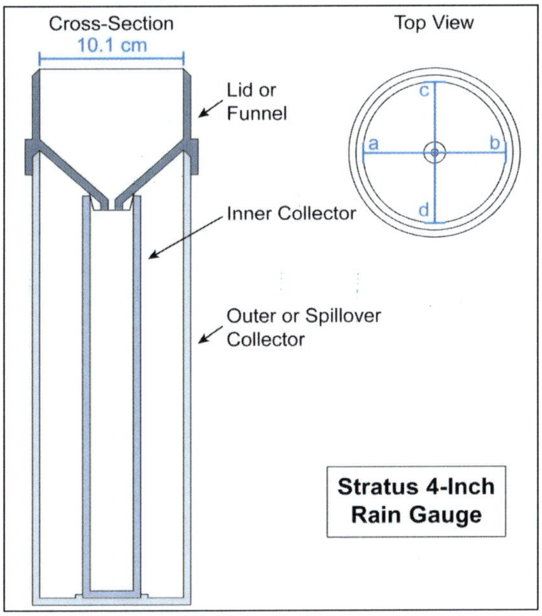

Weighing Bucket Type Rain Gauge.

Weighing bucket type rain gauge is most common self-recording rain gauge. It consists of a receiver bucket supported by a spring or lever balance or some other weighing mechanism. The movement of bucket due to its increasing weight is transmitted to a pen which traces record or some marking on a clock driven chart.

Weighing bucket type rain gauge instrument gives a plot of the accumulated (increased) rainfall values against the elapsed time and the curve so formed is called the mass curve.

Tipping Bucket Type Rain Gauge

Tipping bucket type rain gauge is a 30cm sized circular rain gauge adopted for use by US weather bureau. It has 30cm diameter sharp edged receiver and at the end of the receiver is provided a funnel.

Pair of buckets is pivoted under this funnel in such a manner that when one bucket receives 0.25mm of precipitation (rainfall), it tips discharging its rainfall into the container, bringing the other bucket under the funnel.

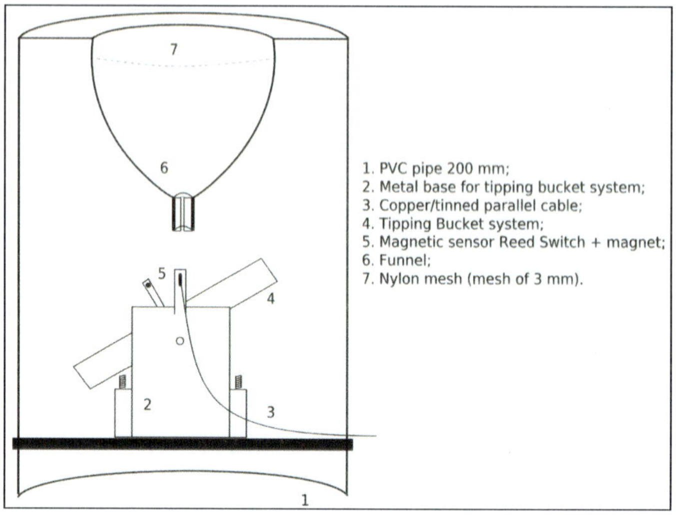

Tipping Bucket Rain Gauge.

Tipping of bucket completes an electric circuit causing the movement of pen to mark on clock driven receiving drum which carries a recorded sheet. These electric pulses generated are recorded at the control room far away from the rain gauge station. This instrument is further suited for digitalizing the output signal.

Floating or Natural Syphon Type Rain Gauge

Natural Syphon or Float Type Rain Gauge.

The working of this type of rain gauge is similar to weighing bucket rain gauge. A funnel receives the water which is collected in a rectangular container. A float is provided at the bottom of container, and this float rises as the water level rises in the container. Its movement being recorded by a pen moving on a recording drum actuated by a clock work.

When water rises, this float reaches to the top floating in water, and then syphon comes into operation and releases the water outwards through the connecting pipe, thus all water in box is drained out.

Weather Balloon

Weather balloon is a balloon used in the measurement and evaluation of mostly upper atmospheric conditions. Information may be gathered during the vertical ascent of the balloon through the atmosphere or during its motions once it has reached a predetermined maximum altitude. Today, atmospheric information is most often gathered by height-finding radar, remote sensing by earth-orbiting or stationary satellites, and aircraft instruments, with weather balloons augmenting the data. Helium, which is less dense than air, is usually used to inflate weather balloons. A pilot balloon is a small balloon (diameter c.1 m/39 in.) whose ascent is followed visually to obtain data for the computation of the speed and direction of winds at different altitudes. A smaller ceiling balloon is used to determine the altitude of cloud bases. A much larger, teardrop-shaped balloon is used to carry a radiosonde aloft. The balloon expands as it rises, usually reaching an altitude of at least 90,000 ft (27,400 m) before it bursts. A small parachute lowers the instruments to the ground. Teardrop-shaped balloons are also used for horizontal sounding of the atmosphere. Atmospheric pressure, temperature, and humidity information may be sent by radio from a balloon; monitoring of its movement provides information about winds at its flight level. Techniques also have been developed whereby many horizontal sounding balloons can be monitored by earth-orbiting satellites that relay information to earth-based stations. The tetroon is a tetrahedral balloon used for horizontal sounding. It was developed to withstand the extremely low pressures of high-altitude flight; the straight seals joining its four triangular faces are stronger than the curved seals of the more traditionally shaped balloons. Tetroons have been used extensively in tracing low-level atmospheric currents by following their movement with radar;

they have thus increased the meteorologist's understanding of atmospheric turbulence, low-level vertical motions, and air pollution dispersion.

Meteorological/Weather Satellite

Meteorological satellites carry sensors that are pointing towards the ground, enabling them to have bird eye view of the globe from the space. There are two types of meteorological satellites characterized by their orbits. They are geostationary satellites and polar-orbiting satellites.

As the name suggests, a geostationary satellite is stationary relative to the earth. That is, it moves above the equator at the same rate as the earth's rotation so that all the time it is above the same geographical area on the earth. In this manner, it is capable of taking cloud images of the same area continuously, 24 hours a day. As it is some 35,800 kilometres from the earth, it is capable of taking cloud pictures covering part of the whole globe.

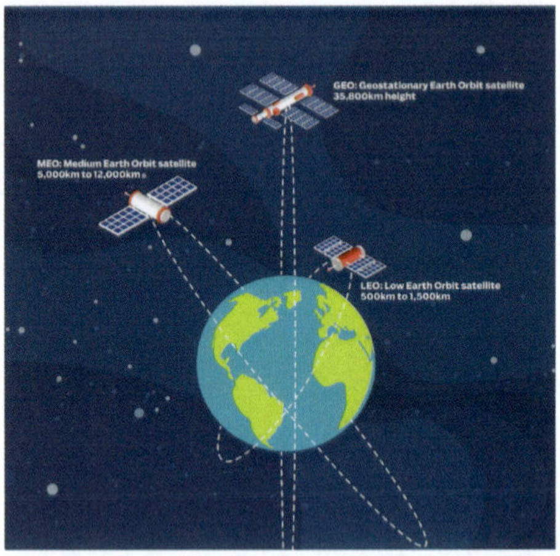

Orbits of geostationary satellites and polar-orbiting satellites.

Polar-orbiting satellites are low-flying satellites circling the earth in a nearly north-south orbit, at several hundred kilometers above the earth. Most of them pass over the same place a couple of times a day. As they operate at a distance closer to the earth, they are only capable of taking cloud images of a limited area of the earth each time. Compared with geostationary satellites, polar-orbiting satellites offer fewer and smaller cloud pictures. However, the advantage is that the cloud pictures obtained are of much higher resolution.

Weather Radar

Weather radar is an instrument used to locate precipitation, calculate its motion, intensity, and determine the type (rain, snow, hail, etc.). Weather radar sends out a pulse of electromagnetic energy into the atmosphere. If that piece of energy interacts with something in the environment, it will reflect energy back to the weather radar and we can sense it. Today most radar use the Doppler effect and have dual-polarization. A Doppler radar determines whether movement is toward or away from the radar. Dual-polarization is a combination of radar design and software algorithms that help identify the type of precipitation.

Radars have four basic components:

- A transmitter, which generates the energy pulse;
- An antenna to send these pulses out into the atmosphere and receive the reflected pulse back;
- A receiver, which detects and amplifies the signal;
- A display system.

Benefits of Weather Radar

Weather radar is an integral part of modern weather observation systems. Weather radar supports operations of:

- National meteorological institutes,
- Hydrological services,
- Aviation authorities,
- Various other applications.

Weather radar enables more efficient decision making, improved situational awareness, better warnings, and forecasts. And when integrated with other information, such as lightning data, weather radars provide an additional layer of meteorological data to increase safety and improve operational efficiency.

References

- Changes-in-the-weather, Changes-in-the-atmosphere, barometric-pressure, barometer: nationalgeographic. org, Retrieved 8 January, 2019
- Ceilometer, technology: britannica.com, Retrieved 15 August, 2019
- Hygrometer: safeopedia.com, Retrieved 13 March, 2019
- Rain-gauge, engineering: sciencedirect.com, Retrieved 20 May, 2019
- Types-of-rain-gauges, water-resources: theconstructor.org, Retrieved 25 July, 2019
- Weather-balloon, meteorology, weather, earth: infoplease.com, Retrieved 5 February, 2019
- What-weather-radar: vaisala.com, Retrieved 19 April, 2019

Permissions

All chapters in this book are published with permission under the Creative Commons Attribution Share Alike License or equivalent. Every chapter published in this book has been scrutinized by our experts. Their significance has been extensively debated. The topics covered herein carry significant information for a comprehensive understanding. They may even be implemented as practical applications or may be referred to as a beginning point for further studies.

We would like to thank the editorial team for lending their expertise to make the book truly unique. They have played a crucial role in the development of this book. Without their invaluable contributions this book wouldn't have been possible. They have made vital efforts to compile up to date information on the varied aspects of this subject to make this book a valuable addition to the collection of many professionals and students.

This book was conceptualized with the vision of imparting up-to-date and integrated information in this field. To ensure the same, a matchless editorial board was set up. Every individual on the board went through rigorous rounds of assessment to prove their worth. After which they invested a large part of their time researching and compiling the most relevant data for our readers.

The editorial board has been involved in producing this book since its inception. They have spent rigorous hours researching and exploring the diverse topics which have resulted in the successful publishing of this book. They have passed on their knowledge of decades through this book. To expedite this challenging task, the publisher supported the team at every step. A small team of assistant editors was also appointed to further simplify the editing procedure and attain best results for the readers.

Apart from the editorial board, the designing team has also invested a significant amount of their time in understanding the subject and creating the most relevant covers. They scrutinized every image to scout for the most suitable representation of the subject and create an appropriate cover for the book.

The publishing team has been an ardent support to the editorial, designing and production team. Their endless efforts to recruit the best for this project, has resulted in the accomplishment of this book. They are a veteran in the field of academics and their pool of knowledge is as vast as their experience in printing. Their expertise and guidance has proved useful at every step. Their uncompromising quality standards have made this book an exceptional effort. Their encouragement from time to time has been an inspiration for everyone.

The publisher and the editorial board hope that this book will prove to be a valuable piece of knowledge for students, practitioners and scholars across the globe.

Index